The Technical Career Navigator

Ray Weiss

Prentice Hall P T R
Englewood Cliffs, New Jersey 07632

Library of Congress Cataloging-in Publication Data

Weiss, Ray

    The technical career navigator / Ray Weiss.

      p.   cm.

    Includes bibliographical references.

    ISBN 0-13-148396-X

    1. Technicians in industry--Vocational guidance. I. Title.

TA158.W445   1995   94-7235

602.3--dc20   CIP

Editorial/production supervision
  and interior design: *Harriet Tellem*
Cover design: *Jeanette Jacobs Design*
Manufacturing manager: *Alexis Heydt*
Acquisitions editors: *Michael Hays*
Editorial assistant: *Kim Intindola / Diane Spina*

©1995 by Prentice Hall P T R
Prentice-Hall, Inc.
A Simon & Schuster Company
Englewood Cliffs, New Jersey 07632

The publisher offers discounts on this book in bulk quantities.
For more information contact:

        Corporate Sales Department
        P T R Prentice Hall
        113 Sylvan Avenue
        Englewood Cliffs, New Jersey 07632

        Phone: 201-592-2863
        FAX: 201-592-2249

Printed in the United States of America
  10 9 8 7 6 5 4 3 2 1

ISBN 0-13-148396-X

Prentice-Hall International (UK) Limited, *London*
Prentice-Hall of Australia Pty. Limited, *Sydney*
Prentice-Hall Canada Inc., *Toronto*
Prentice-Hall Hispanoamericana, S.A., *Mexico*
Prentice-Hall of India Private Limited, *New Delhi*
Prentice-Hall of Japan, Inc., *Tokyo*
Simon & Schuster Asia Pte. Ltd., *Singapore*
Editora Prentice-Hall do Brasil, Ltda., *Rio de Janeiro*

Dedication

To my wife Colette

She's put up with a lot over the years (I finally did fix the Saab that lay disemboweled for 2 years in the driveway). Bless her.

To Eph Konigsberg

Friend, literary critic, and president of Konigsberg Instruments (Pasadena, CA). He took time out from design and running a company to critique and help shape this text (He's tough minded; he even edits cereal boxes at breakfast). Thanx.

To the Los Angeles County jury system

For the impetus to do this book. Like many technical people I've never had time for jury duty. This year in a fit of remorse I went on jury selection for a murder trial, even arranging to work nights during the trial. Needless to say, I was booted off the jury (if you really want to be on a jury — be meek, don't act technical, stare at the floor, and mumble a lot). So instead of jury duty I wrote this book. Thanks, but no thanks.

Acknowledgment

Back in the early 1970s Robert Townsend wrote *Up the Organization*, a classic text that defined the corporate mores, management techniques, and structures that characterize successful corporations of the 1990s. It took over 20 years for Townsend's message to sink in and permeate corporate America.

Townsend, the CEO of a progressive corporation (AVIS — "We're number two, but we try harder"),— showed how to empower management and employees to build a successful, slimmed-down modern business. *Up the Organization* was an immediate best seller and a revised version, *Further Up the Organization* remains in print as a paperback. Still relevant, the book is a classic of clear thinking and incorporates a succinct writing style and form. Buy it and read it — you won't be disappointed.

Imitation is said to be the sincerest form of flattery: this book builds on the form, structure, and hopefully, spirit of *Up the Organization*. However, I am not a Robert Townsend, instead I'm an engineer, programmer, ex-manager, and technical editor, a veteran of 20 some years in engineering and software. This book is a distillation of that and others' experiences.

You have a choice: you can learn from your own experience, or from the experience of others. Learning firsthand can be expensive, and even painful — it can be, as they say, hard on the people and the furniture.

Have fun,

ray weiss

Contents

30 Job Search Tips

Introduction

TO: The Reader

FROM: The Author

SUBJECT: Surviving and thriving in the high-tech 90s

Experience is the worst teacher; it gives the test before presenting the lesson.
— Vernon Law, ex-pitcher, Pittsburgh Pirates

Technology is no longer a sure thing. We now work in a highly competitive and unstable world. Today's successful companies can easily end up as tomorrow's Chapter 11 casualties. Thus, job security no longer lies in who you work for, but rather in what you can do.

It's up to you, not your management, to ensure that you build up skills and capabilities. It's now part of your job to build a better technical you, not just build products. Twenty years ago it was possible to graduate from college, walk into a major technical employer, and stay 20 years for a safe, secure career. No more.

Let's face it, playing with technology is fun, that's why we're in this business. In technology-driven products there are hard results, there are successes (and failures). But you can see, feel, and even taste the results of what you do. Unlike many jobs and professions, there are hard milestones and real wins.

Conversely, dealing with people and organizational problems is a chore for most technical folk. Most of us went into technical careers because we preferred to deal with things and concepts rather than people problems. But in today's superheated work environments, solving technical problems may not be enough.

Technical professionals can no longer afford to let their managers, their companies, and fate determine their futures. You can survive and even thrive in this new technical world, but there's a price — consciousness. Drifting along won't cut it, unless you're comfortable with ending up selling insurance or flipping burgers.

Read this book. Gathered here, culled from many sources, are insights, curses, and band-aids that can help. There are over 100 items, ordered alphabetically and crammed into one or two page articles. Start anywhere and use what you can. It may make the critical difference.

Attack Memos

Let us work without disputing. It is the only way to render life tolerable.
— Voltaire (1694 – 1778)

It's easy to be brave from a distance.
— Aesop (620 – 560 BC)

Don't write attack memos. Nothing destroys a working environment faster than interchanges of memos criticizing people or organizations.

1. One manager stopped attack memos deadbang. It was simple: if you wrote a memo attacking someone or a group, you ended up on his carpet, memo in hand, reading it face-to-face to the criticized party.

   Once in place, that rule virtually dried up attack memos like a drought in Texas. To write one, you had to be very, very sure of your ground as well as be willing to go to the mat in defending it. While this rule discouraged frivolous attack memos, it also insured that legitimate concerns got immediate, face-to-face attention.

2. It's surprising just where yesterday's written criticism can later surface and cause downstream problems.

Nothing stays secret (even Nixon's tapes were opened to scrutiny). Moreover, innocent criticism can be easily misconstrued, or worse, be used by others for political ends. So, even if your boss asks you to write a critique of someone, avoid it. Do so only for critical matters, such as project life and death.

3. Attack memos are inherently unfair, and generally give an unbalanced account. If you must write a criticism, go see the party under scrutiny first, and get their side of things. They may be able to clean up the problem right then and there, and eliminate any need for management interference (and your memo). If not, be sure to include their side of things in your memo.

4. If you do write an attack memo, keep it on a high level, and keep it clean. Attack a problem — not a person — and present a solution. Don't, repeat don't, get personal. And above all make absolutely 100 percent sure that you're right: the stakes are too high, both morally and politically, to be wrong.

5. A good curb on getting carried away with an attack memo is to write nothing that you would not be willing to read face-to-face to the criticized party.

6. Today's E-mail ups the ante for attack memos. The temptation is always there. Feel frustrated? Just dash out a critical memo and feel better. There's no built-in delay to give you time for second thoughts; you can instantaneously react and launch an attack. Not only will the addressees get it, but that memo will float in the E-mail system limbo, just waiting for exhumation.

Aural Versus Visual

The journey of a thousand miles must begin with a single step.
— Lao Tsu (604? – 531 B.C.)

Listen and learn.
— Franklin D. Roosevelt

Don't waste time presenting your ideas in the wrong form. Find the primary presentation approach that your manager(s) prefer and use it. Match your presentation to the preferred input mode rather than relying on those methods that you're most comfortable with.

Managers generally are either aural, gathering data by hearing it presented, or visual, absorbing information by reading it. Be flexible. Some managers prefer one mode to screen topics, followed by the other for detailed analysis, while others rely on reading to absorb basic data and proposals, but use interviews to set the data into perspective and resolve details. Be patient; it can take time for your views to prevail; some managers act quickly, while others need a long gestation time before giving birth to action.

Bad Assumptions

We sit on our assumptions.
— Anonymous

It ain't what we don't know that gives us trouble, it's what we know that ain't so.
— Will Rogers

People who don't question the assumptions made going into a problem often end up solving the wrong problem.
— Marcian E. (Ted) Hoff, Jr., coinventor of the microprocessor**

Always check your initial assumptions. Nine times out of ten, most major design problems are caused by incorrect assumptions made in the early design stages. Just as a good driver always looks back before changing lanes, a good designer automatically checks his or her initial assumptions (this too can avoid unexpected collisions).

1. Go back over the original specs and see if you've created an arbitrarily bounded solution. The classic example of an unbounded solution can be seen in the 9-dot puzzle with 9 dots arranged in three rows of three each. You're to connect the dots using only four linked straight lines. An easy solution is to violate the invisible boundary around the 9 dots — going outside the bounds.

2. Bad assumptions can add unnecessary complexity. A clean design is usually the result of reducing a problem

5

to a minimum requirements set and then creating an optimum solution. It's not an accident that most military systems are so cumbersome and ungainly. Instead of aiming for a minimal solution, the government project management protects itself by forcing designers to conform to elephantine specifications.

3. To minimize the damage from implicit assumptions in your design, make a list of the assumptions that you have made about the design and its requirements. Then supplement that list with another that details the counter-assumptions — the things that are not true, that you have eliminated from the design specs.

4. Incorrect initial assumptions limit software use. One reason that new software tools are so difficult to use effectively is that most of us carry a set of assumptions with us on how a package should work (like the old one). It takes us time to reeducate ourselves, to discard wrong assumptions. Application engineers, who support complex software, spend much of their time correcting initial user assumptions.

5. Most product documentation is not explicit enough to stop user/designers from going off half-cocked with erroneous ideas about the product and how it works. Most documentation tries to explain what something is, rather than what it is not. Very few users tackle a software tool with an open mind (see Documentation Daze).

6. Good programmers and hardware engineers (logic designers) tend toward the paranoid. They don't trust any data or hardware signals that come their way — they protect their designs from external corruption from bad inputs or misassumptions about

them. They make no assumptions about the validity of incoming data and check everything.

***Career / Profiles: Marcian E. (Ted) Hoff, Jr.,* by Tekla S. Perry, Feb. 1994. *IEEE Spectrum Magazine*, IEEE, NY, NY.

Bad Drives Out Good

Bad money drives out the good.
— Sir Thomas Gresham (1571 – 1579)
(Gresham's Law — economics)

It's common knowledge that bad management will drive out good people. The opposite is also true: bad employees can drive out good managers.

If you had managers who made your life miserable, sooner or later you would leave. Managers are no different; they can only take so much before they too will move on to more bucolic pastures. Just as an inflexible or dictatorial manager can drive you out the door, so can bad or troublesome employees drive out good managers.

So, if you have good managers, treasure them. Even better, protect them. Don't let them be driven out by problem employees. And while you're at it, take a look at what you are doing as well. Are you making life difficult for your managers? Are you continually using them as a complaint sounding board? If so, maybe you should cut them some slack and back off.

No one likes a steady diet of bad news and difficult decisions, and that includes managers. Don't just bombard your managers with the distaff side of things; try to share some triumphs and good news with them as well. They'll appreciate it. They like to win too.

Your management, especially your direct superior, is the best resource you have. Protect them.

Be Noticed

Concealed talent brings no reputation.
— Desiderius Erasmus (1466? – 1536)

In the 19th century, Bishop Berkeley asked if a tree falls in the forest and no one hears it does it make a sound? In effect, did it happen? (Does the refrigerator light stay on when the door is closed?) Unfortunately, the same can be asked about work that you do. If no one notices, did it happen? Does it matter?

Most technical people simply want the maneuvering room and resources to get things done. Few like office politics or spending time in self-promotion. Unfortunately, it's not enough to get the job done. You must also make sure that your management knows your accomplishments. This tactic is not shameless self-promotion — it's simple prudence.

A few years back, while stalling on a large task, I went on a spree of reading World War II novels, mainly about the professional military. I noticed an interesting trend in these stories — that there were basically two kinds of officers

11

portrayed: those who worked at succeeding at the current mission, and those who concentrated on getting promoted. Well, the missions got done, but the self-promoters got ahead, rising to where they could eventually do some real damage.

Sound familiar? Don't give self-promoters the advantage. If they succeed, they'll end up running things, but without the base experience at making things work — they were too busy getting promoted to pay attention. So talk to your managers. If you want to succeed, be more focused; try to understand how your contribution can aid the overall business objective. How can you better your design and skills to help meet those goals?

Performance and results come first — always. Put your back into your tasks and do what needs doing. Just make sure management knows what you've done and what you're capable of doing.

Betting on Instinct

*No great marketing decisions have ever been made
on quantitative data.*
— John Scully, Apple Computer

Often you have to rely on your own intuition.
— Bill Gates, Microsoft

Taking on a new product or new product direction really constitutes a bet on one person's product instincts.

Most organizations have comprehensive, detailed product marketing requirements. A new product or a new product direction must run a marketing analysis gauntlet before management commits to building and marketing it.

Yet, as most experts know in their heart of hearts, most marketing statistics aren't worth the paper they are printed on. Aside from defining marketing and distribution strategies, most premarketing plans are a waste of time and resources.

A company really bets on the market savvy and product instincts of that new product's champion. If he or she is in tune with the market and customer needs, if their instincts are on target, then the product may be a winner. Thus, a marketing study is really an exercise to educate the product champion(s), to make sure that they consider all factors. What it is not, however, is a comprehensive proof of product concept for the marketing committee.

Books and Winners

*The most technologically efficient machine that man
has ever invented is the book.*
— Northrop Frye

Books and winners have a chicken-and-egg relationship: the
winning technology products generally have books covering
them. Which came first? The books or product success? It
doesn't matter. Successful technical products, both hardware
and software, now have commercial documentation. And the
more successful they are, the more books they have.

Today's commercial documentation, computer books and
many engineering texts, are far better than yesteryear's documentation. A PC or MAC hacker has better documentation
at his or her fingertips these days than any mainframe programmer had in the days of yore. These off-the-rack books
are readable and try to bring technical mastery to a wide
range of users.

Technical product books are not just the product of
fringe designers or journalists. Mainstream hardware and
software vendors are turning to book publishers for documentation. Instead of pushing out more and more manuals,

they are publishing them as books and turning distribution over to bookstores and publishing houses.

So if you, your group, or your company are rolling out a new hardware or software product, and if you want to hit the big time, consider doing commercial, book-type documentation. Such a book becomes a continual advertisement. It also makes a great sales tool — something valuable to leave with customers.

Doing a book on your product may not be as difficult a task as you imagine. After all, you've been living with the product over the development cycle and most of the data already exists as raw documentation. Moreover, having done a book can't hurt your career; it marks you as someone who can organize and write.

Boredom

Better to wear out than rust out.
— Bishop Richard Cumberland (1631 – 1718)

When you're green, you're growing. When you're ripe, you rot.
— Ray Kroc, founder, McDonald's

Boredom is the first sign that you're in trouble. If you're bored, you are either not working up to your potential or not building new skills.

Change something. Get new work in your group; find something interesting to do that helps both you and your group; or go to a new group in your company; or get a new job.

The most valuable thing you possess (other than integrity) is your professional and technical skills. Don't let them rust or wither away. Over the long run, you are paid, either as a manager or technical contributor, for what you can do. The less you can do, the less value you have.

Besides, who wants to be bored? Work can be just as engrossing, as challenging, and as much fun as you make it. Find some good stuff to do and do it. Take the risk. It's a bigger risk to do nothing.

Bottlenecks

A bottleneck is any resource whose capacity is EQUAL TO or LESS THAN the demand placed on it. And a non-bottleneck is any resource whose capacity is greater than the demand placed on it.
— Eliyahu M. Goldratt, Jeff Cox
(The Goal)

Bottlenecks are not just confined to the factory floor, they are everywhere. You may be more constrained by built-in bottlenecks in your organization than you realize. Look for bottlenecks and learn to schedule around them to increase both your and your organization's efficiencies.

In every process there may be one or more bottlenecks, constrictions that slow or limit the flow of results. The trick is to identify these bottlenecks and then figure out how to minimize their effects by working around them, or at least pacing work to optimize flow through them.

You can get a good lesson in bottlenecks and how to deal with them in *The Goal*, a book by Eliyahu M. Goldratt and Jeff Cox.** Intended as an alternative teaching mechanism for a new method of production control, *The Goal* is the only novel on manufacturing that I know of.

Not only that, but *The Goal* is also a good read. And it's useful too; a number of companies use the book to make managers aware of production scheduling problems. For example, it's required reading at International Rectifier, the leading U.S. power MOSFET supplier.

One point *The Goal* makes is that you should leave some slack in your system; it's almost impossible to run resources at full capacity. Another is that it's easy to spot bottlenecks: just look for piles of pending work queued up for processing. The overloaded resource in front of the pile is a bottleneck. Check around your organization (I did); you may be very surprised at the number of built-in bottlenecks that you find.

** An underground classic, *The Goal* has sold over 800,000 copies since publication in 1984. The book is unique in another way: there are at least three versions; Goldratt and Cox keep rewriting and expanding the ending to strengthen the book's manufacturing lessons. There may be as many as four distinct endings to the novel. All are good.

Breaking Owsie Chains

What have your done for me lately?
— Anonymous manager

Owsie Chains, as in "I Owe You" chains, can put sand in the organizational gears. Over the years, organizations build up these Owsie Chains, which end up constraining what managers can ask or expect from subordinates.

Suppose I work for you. And a while back, I saved your bacon. On one of those nightmare projects I put in 80+ hours a week for six months and made the project a success. Well, you owe me for that. Over time, these debts can build up into a series of Owsie Chains, chains that restrict what you and other managers can require from those who work for you.

Worse, Owsie Chains give subordinates a false picture of security. Owsie Chains can lead them into believing that past achievement, rather than current performance is enough to ensure their jobs. Many companies boost efficiencies by simply moving managers around, thus breaking these Owsie Chains. The newly installed managers simply ask: "What can you do for me now?"

Technical design and management are fast-paced and not a place to stop and smell the flowers. Owsie Chains are potential traps because they can lull individuals into slacking off and loosing the edge needed to survive.

So don't live on past glories or count on Owsie Chains for security. Continue to grow and do. Let what you've accomplished be a prologue to what you will do in the future.

Burnout

Happiness is neither virtue nor pleasure nor this thing nor that, but simply growth. We are happy when we are growing.
— William Butler Yeats (1865 – 1939)

In reality, killing time is only the name for another of the multifarious ways by which time kills us.
— Sir Osbert Sitwell

Watch out for burnout. Monitor yourself and change as needed to avoid burning out. Generally, if you can't be effective, then you're in danger of burnout.

The causes of burnout are many; the cures, surprisingly simple. Causes can include: boring, repetitive grunt work; no-win, toxic office politics; head-in-the-sand management, which ignores critical problems; ineffective team members; bad planning; and impossible project schedules and promises.

The cures are straightforward: get work that is challenging; stay out of office politics; put your energies into your project segment and let others reap the bitter harvest of bad implementations; get some authority to help weak team members; or, if worse comes to worse, get another job.

Your security, your industry worth is based on what you, as a professional, can do. You cannot allow any company, project, manager or coworker to affect you to the point that you lose your effectiveness. Take whatever steps are needed to protect your core competence.

By the You-Know-Whats

*If you have them by the "you-know-whats" their hearts
and minds will follow.*
— Chuck Colson

Get people to use your hardware or software products, and
they'll be tied to them.

Back in the Vietnam War days, then-president Nixon
had Chuck Colson working for him. Colson, a real hatchet
man, was known for his ruthlessness in serving Nixon. He
was later convicted and imprisoned in the Watergate scan-
dal.

Colson was infamous for his noxious statement on paci-
fying Vietnam villagers (see above). Here's the corollary for
the hardware and software business: "If you can get them to
use your hardware or software you have them by the 'you-
know-whats'."

Why? Because once someone pays the up-front time
costs of learning a software package or designing-in a micro-
controller, they don't want to pay that price again for a rival

product (which explains why vi and Wordstar, early and hard-to-use word processors, are still around).

A number of vendors use a variation on this condition: they market low-cost, easy-entry products to entice and lock in new customers. In "market speak" this tactic is called "lowering barriers to entry and leveraging your installed base."

Can't Fool People

We are never deceived; we deceive ourselves.
— Goethe (1749 – 1832)

Truth sells itself; lies deed your soul to the Devil.
— Rear Admiral Dave Oliver, Jr., U.S. Navy

Generally, you can't fool people, unless they're willing to be fooled, usually because they don't care or are not paying attention. However, once they do care and pay attention — look out!

So don't try to fool anybody. It is far, far better to just play things straight. It's easier and it"s also far more effective.

I once worked for a company that was a subcontractor to a larger, prime contractor. We were working on a large contract and this major contractor started lying to the customer. Things got so complicated that they actually kept a log book with all their lies in it. Why? So that they could track what they had said and not trip themselves up. I quit in disgust.

Lies are funny things; they can get complicated. Whereas the truth is much simpler and far easier to remember.

Fooling people may take care of today's sticky situation, but it sets you up for tomorrow's fall. Retribution delayed is usually retribution doubled. So if you have to take a hit on something you did, take it up front — it's simpler and far less painful.

Change

*Rapid, drastic change means the intrusion of the future
into the present with the result that the present has
become as unpredictable as the future.*
— Eric Hoffer

*When choosing between two evils, I always like to try
theone I've never tried before.*
— Mae West

Change — count on it. The one thing that most technical
managers, engineers, and programmers can be sure of is
massive, continuous change. If you want a guarantee of stability and slow change, take up cosmology or geology. Everything else, especially the technical world, is subject to
change:

1. Your boss. Never count on having the same manager
   or management hierarchy. Management shakeups,
   buyouts, layoffs, expansions, mergers, retirements,
   leaving for a new job — all can lead to new management.

2. Your job. It can go away, can change. You can shift to
   new, different work.

3. Your company. It can merge with another, split up,
   change directions, vanish.

4. Your technical specialty. It can become obsolescent, shift to a new technology, become more complex and software intensive. It can be overshadowed by a competing specialty.

5. The economy. It can improve, hit a boom period, or drop like a rock into major recession.

The bottom line: nothing in our work world is really stable. Yet most of us continue to think of tomorrow in terms of today — we should know better. Technology drives much of what tomorrow can be. Who could have predicted the incredible rise of personal computers, or the microprocessor revolution, or the wide sweep of software today?

There are no guarantees except one: new technology needs technical folk for its creation, deployment, care and feeding. Be there.

Clusters

Hierarchy is dying. Everyone is sick of the rituals, delays and inefficiency. It's almost a corpse and soon will be buried.
— D. Quinn Mills, Fortune 500 staffer
  (Rebirth of the Corporation)

New organizational forms are replacing the traditional top-down, management hierarchy. Clusters — relatively self-sufficient, mini-organizations — are gaining popularity as an alternative to rigid hierarchies.

Traditional top-down, hierarchical management evolved from early military structural forms. Our current organizations have more immediate roots in shipping, and particularly railroads, which pioneered modern management structures.

These traditional top-down structures form a pyramid, made up of mini-pyramids, each headed by a manager controlling a number of managers or organizations below. In this structure, managers typically have 3 to 7 lower level managers reporting to them. This number is referred to as their span of control. (Old rule for span of control: 3 or less, you're lazy; more than 7, you're crazy.)

With clusters, management takes a different tack in forming mini-organizations, each targeting a specific business sector or niche. Cluster internal structures are not hierarchical; instead individuals and groups are encouraged to take charge of their areas of expertise. Relatively self-governing teams are formed. These teams have to meet business and organization goals, rather than be closely monitored by a next-level manager.

Clusters still have managers, albeit fewer than a like size organization would have had in days gone by. These managers act more as facilitators and monitors than traditional hands-on management. Consequentially, they have a much larger span of control. Day-to-day decisions are made as low as possible in the organization, at the group or individual levels in the clusters. By placing responsibility at a low level, organizations can react much faster as market conditions and customer needs change.

In clusters, the organizational meat and muscle is out there in clusters that directly interact with customers. Top management and corporation-level services are concentrated in central services that serve the clusters.

Engineers, programmers, and other technical professionals will be winners in this shift to cluster organizations. Many already work in small to large project teams. Such cluster organizations optimize these project structures, eliminating the current hard-to-work-with rigid management levels. Cluster organizations tend toward flattened, shallow structures with much faster access to top management levels. At the same time, the clusters will be more autonomous: technical people in the clusters will have more autonomy and product design responsibility

Clutter

Where is it?
  — All of us

Clutter defeats us all. There's nothing more frustrating than not finding important information when you really need it.

There are clean-desk, organized folks, and there are the rest of us. Out of school, I worked for Howard W., a senior project engineer who had — and I'm not making this up — stacks of paper, a foot high, covering his desk. He could find anything he needed in it, down to the right pile and paper depth (he'd gently extract it, use it and then just as carefully return it).

I followed in his path: in most jobs I never had more free space on my desk than could accommodate my forearms and a sheet of paper. I probably quit and took other jobs just to get a clear working space. As a technical magazine editor, my clutter has gotten even worse; the depth of each week's mail pile can be measured in feet.

But I've finally defeated clutter; I now work on a clear desk — I can see bare wood. It was easy: I have two offices!

One is an outside office, where I have piles and piles of stuff, all moldering in drifts of paper awaiting attention. And I have an inside office, where I put only those things that I am directly working on.

It works! No clutter. Even better, I no longer have the guilt piles of stuff that I would put on my desk to remind me of things to do. The bad vibes are gone; I now can concentrate on what I am doing.

Try out this anticlutter tactic. It'll work in an office or even a cubicle. Just put all the scheduling stuff, the guilt piles and pending work on one side of your space. And on your desk side, put only the work in progress. You'll be surprised at how virtuous and pristine pure you'll feel, working on a clean desk.

Here's a clutter story: as an editor writing an article, I actually misplaced and had to request the same material from one company FOUR times. Each time I got the material it would just sink out of sight in the surrounding clutter. Around the fourth request it got pretty embarrassing to ask for another set of data. Just what can you say?

Coffee Test

*A leader is not an administrator who loves to run others,
but someone who carries water for his people so that they
can get on with their jobs.*
— Robert Townsend, AVIS

Here's an interesting test of managerial reasonableness: the
"Coffee Test." It's simple — does the manager make coffee
when the pot is empty? Or does he or she always leave their
cup and wait for someone else (read underling) to do it?

Good managers resemble football or basketball coaches;
they do what has to be done for team success. If that means
making coffee or putting team members at ease, they do it.
On the other hand, Theory X ** managers who try to rule by
fiat or by fear are best avoided. It's difficult for them to build
good teams, for in their minds they're the star player. Unfortunately, one player, no matter how good, can't substitute for
an entire team.

I once worked at a small system software house. The
president always made coffee; the executive VP never did: he
thought his time was too valuable. The president always listened and tried to do the right thing; the VP rarely listened

and always tried to do the political thing. Ultimately, he almost buried the company.

Use some common sense with this test. Some managers are truly busy and don't have a lot of time as they run from pillar to post, to take time to make coffee. This test is telling for those managers who have the time to do it, but choose not to based on some idea of hierarchical importance.

** *The Human Side of Enterprise*, by Douglas McGregor, 1960. In this classic text McGregor defined managers as falling into two classes: Theory X and Theory Y managers. Theory X managers followed the top-down, don't trust the workers method, while Theory Y managers aimed at empowering their workers.

Company Personality

That's the spirit that brought us fame!
We're big, but bigger we will be.
We can't fail for all can see.
That to serve humanity has been our aim.
Our products are now known in every zone.
Our reputation sparkles like a gem.
We've fought our way through, and new
Fields we're sure to conquer too.
Forever onward IBM.
  — Old IBM company song

People, dogs, and cats have personalities; so do companies. Each company develops its own unique personality, its ways of working and doing business. So try to work at a company whose personality meshes with yours.

Company founders and plank owners (first employees) create a company personality, a way of looking at the world, doing things, and working with customers. This personality outlasts the people who started the company. It's similar to the effects of a freeway accident: hours after an accident is cleared up, traffic still snarls, as if held by the memory of the accident.

Company personalities are known industry wide. Some companies are market oriented; others are technology driven; some are calm places to work; others thrive on confrontation. All things being equal, try to work at a company that matches your way of working, although a new environment can help to stir up one's creative juices and help your professional growth.

Complacency

If you are doing something the same way you have been doing it for ten years, then chances are you are doing it wrong.
— Charles Kettering, GM Technologist

Complacency: something you can't afford any more. The days of leisurely product development and casual technology tinkering are over. Fast product turnaround, technology changeovers, and shifting markets killed it. These days it pays to be continually on the search for product and technology advantages, as well as new skills. Yesteryear's market verities and technical skills may just not be valid any more.

Complacency is something that your company can't afford either. In today's churning markets a complacent company already has one foot on the banana peel of failure. Not paying attention and failing to react quickly to market conditions can cost companies dearly. It's terribly easy to snatch defeat from the jaws of victory; the bone yard is full of careless companies who did.

If your company is too complacent; if it believes that its past defines the future; if it is unwilling to change to stay competitive — you have a problem. You have to ask yourself some tough questions: Is there enough time to change? Is management willing to try? Can they survive? If the answers read like a cup of bad tea leaves, then perhaps it's best to find another job or at least find a growth area within the company.

Conceptual Maps

*Mass and friction may seem less formidable concepts when
presented in terms of an elephant sliding down a grassy hill.*
— Sir Arthur Eddington

Conceptual maps are needed for documentation. The problem with most documentation, both for hardware and software, is that it concentrates on details and textual presentations, not overall concepts.

More than a few years ago I drove from California to Pennsylvania to work on operating systems at Sperry Univac (now Unisys). Driving from the West Coast I arrived in Blue Bell and promptly got off the turnpike the wrong way. In the almost two years that I was there, I never got North straight. I was always getting lost. I had started off with the wrong map and never recovered.

I suspect that's exactly what happens with many technical products. Users start off with the wrong assumptions and have a hard time ever getting their bearings. So I suggest that most products present large, graphical concept maps in their documentation. Not only can users start right off, and not wander off with wrong assumptions, but product support will be a lot cheaper (an ounce of prevention is worth a pound of repair — or a few pages of conceptual maps are worth manuals of detail).

Concurrent Engineering

The 6 P's: Poor planning produces piss-poor performance.
— U.S. Marine Corps proverb

Concurrent engineering is the latest engineering fad. Many companies already tool up their products using parallel rather than sequentially staged development. Aerospace and Fortune 500 companies increasingly count on concurrent engineering to speed up development cycles.

Close cooperation between different departments is now possible with today's development tools, which can share a common database. Thus, designers can do a trial board placement and layout while engineers finalize the board's logic, other engineers finalize the mechanical details, all while the programmers churn out the code.

There is, however, a danger in this tactic if management tailors their concurrent engineering tools to match their organizational structure. Why? Because most companies really have two organizational structures: a formal one defined by organization charts and such, and the informal structure that reflects the paths people follow to get things done. Thus, if you build tools that mirror the official organizational structure and that structure doesn't parallel the working one, you may inadvertently build gross inefficiencies into your concurrent engineering setup.

Confrontations

*To argue with a man who has renounced the use and author-
ity of reason is like administering medicine to the dead.*
  — Thomas Paine (1737 – 1809)

Best avoided.** If you see an ugly, irrational confrontation
coming, pick the time and place — the ground where you can
win, when you are prepared to make your case, and (hopefully) your opponent is not.

Sometimes a nasty clash with an individual on some
issue is inevitable. If that's the case, pick the occasion where
you're ready and your opponent is off-guard. (Note: Never do
this unless it's a critical, high-stakes situation and your
opponent deserves such harsh treatment.)

Make absolutely sure that you are in the right. Don't get
personal, and keep your discussion at a high level, in terms
of what's good for the project, organization, or company. And
when you do win, be gracious. Do what has to be done. But
try not to humiliate your opponent; let him or her be part of
a common solution — let them save face. Remember, the goal
is not to crush your opponent but to get the things done that
need doing.

However, if you and your opponent have differing views but can both be rational about resolving the differences, then set up a meeting where the alternatives can be presented and honestly evaluated. Get a level playing field and let the best proposal win.

**Unless you work for a company that encourages confrontation to foment and solidify ideas (like Intel), in which case you'd better get good at it.

Consent
of the Governed

You do not lead by hitting people over the head — that's
assault, not leadership.
— Dwight D. Eisenhower

Managers rule with the consent of the governed, their subordinates. Managers do not, as some believe, have a Divine Right to rule. If you are a manager, take care of your people, or they"ll revolt, and like Louis IV, you could end up a head shorter.

What do you think happens when a group refuses to follow its manager's lead? After all, what's more replaceable, the manager or a design team? I've seen a few revolts; and in just about all cases, guess who lost? The manager. In one instance, a software manager had a terrible temper and his programming team became fed up with his tantrums. Of the 10 people reporting to him, five quit in a single day: the next day he was a programmer.

In another case, a hard-charger took over a development group and made a "stand-up and be counted or leave" speech. Enough people stood up and went to the VP and wanted to leave, that he was told to pull in his horns. He eventually left the group.

As a manager, if you support your people, they will support you in return. Catch their mistakes in time and they will return the favor. Without that quid pro quo, sooner or later you will make a mistake and they will let you suffer the consequences.

Consulting

If you pay peanuts you get monkeys.
— Anonymous

Consulting is an increasing option for many technical people.

1. Flattened organizations are hiring expertise directly, bringing in consultants for occasional needs rather than keeping people on staff. Additionally, many companies are turning to their suppliers to provide key design technology, rather than keeping it in-house. These trends represent solid opportunities for technical consulting.

2. To be a successful technical consultant you must keep the pipeline filled with work. Thus, while you're finishing up one contract, you better be out there prospecting for the next.

3. In setting your rates, remember that you have a higher overhead than you had working direct. You need to pay for insurance and other perks that are normally furnished to direct employees.

4. Many engineers become frustrated working for a company, and turn to consulting, reasoning that if they must work at something they don't like, they will at least be well paid for it. Ironically, by becoming consultants they end up doing work that makes them even more unhappy. Many companies bring in so-called consultants, actually line engineers, to do the grunt work that the regular employees don't want to do (they too want more interesting, challenging work). After a few years of this kind of work, a consultant can become virtually unemployable.

5. Make your nut. If consulting is a path you want to take, you should calculate your nut, the amount of money you must make each week. Use that number to set your rates. The first lesson in consulting is not to give your work away — you must make that nut, week in and week out.

6. Remember, you must continue to grow your skills. If consulting assignments don't expose you to newer technology and design methods, then you must invest your own time to keep current.

7. Cash flow. To survive as a consultant you need a stable cash flow to cover expenses and your salary. Factor into your cash flow projections the time it takes for your customers to process payment for your services. Many firms, particularly the larger ones, deliberately age their accounts payable. You might consider adding a 2% discount for fast payment (20 or 30 days) to put your payables on a fast payment track. Talk to your customers' accounts payable folks and find out how their system works.

8. Many companies have a direct need for top-level design consultants. They cannot keep the talent in-house because there is not enough work to keep a top designer busy, but they still need the design exper-

tise for key projects. You can build quite a practice supplying that expertise. Just be sure that you have it.

9. Consulting can take three forms: as a top-level consultant or architectural adviser, as contracted labor, or as a subcontractor with a contracted technical result or product. All are basically consultants. The first are highly paid (and listened to) experts; the second are basic technical labor similar to in-house designers; and the third are developers with a level of product independence and commitment.

Creativity
Is Not Enough

Ideas are cheap.
— Anonymous

It's a long road from a creative idea to the first dollar of sales.
— Aart de Geus, Synopsys

Creativity is only a down payment, a start. You must take a creative idea and make it reality. Talking about it is not enough — you have to implement it.

Ideas are cheap. Just fill an office with five engineers or programmers and you can pump out maybe 50 to 100 new ideas an hour. No sweat.

Many technical professionals think that an innovative idea is its own currency. All they have to do is think it up, and then management will take over and run with it. Fat chance. To most managers, the true test of an innovation or idea is how far its creator is willing to take it. To them, an idea without implementation, or at least an implementation plan or strategy, has little immediate value.

So pick your innovations carefully. Remember an idea is not enough; you have to be prepared to make it real.

53

Credit Theory
of Management

True leadership must be for the benefit of the followers, not the enrichment of the leaders. In combat, the officers eat last.
— Robert Townsend, AVIS

It's not enough as a manager to tell your people to do something; it's far better if they want to do it, if they want to help.

Management can be compared to credit banking. You build up an account with each employee: you do the little things that help keep them focused; you get them the resources they need; and you fight their battles for them — building up credit. Then when the time comes that you need things done, you draw on that credit to get them done. Simple.

And it's effective. On one level, you are removing the distractions and barriers that limit your people's ability to produce results. Yet on another level, you're building up credit to draw on as needed. Everybody wins.

Decisions

Decisions are easy; it's the things that follow that are hard.
— Anonymous

The easiest thing of all is to deceive oneself: for what a man wishes, he generally believes to be true.
— Demosthenes (384 – 322 B. C.)

Most changes of direction in our technical lives are not the result of dramatic major decisions, but rather accrue from many minor decisions.

Only a few times in life do we actually sit down and make a major decision that turns our technical or personal lives onto a new course. Instead, we tend to make lots of small, seemingly minor decisions, that each shift our course a few degrees or so. However, cumulatively, these changes can end up drastically changing course with shifts of 90 degrees or more.

There's nothing wrong with this incremental decision process. But you should be aware of its cumulative long-term consequences. It's best to hold an informal, where-the-heck-am-I? review every year or so just to see if you're headed in the right direction. You might be surprised at what you find.

Design Reviews

The biggest problem in the world could have been solved when it was small.
— Lao Tsu (604? – 531 B.C.)

Every man is his own best critic.
— William Somerset Maugham (1874 – 1965)

Getting it right the first time is increasingly important. Design reviews help you get there.

The major benefit from most design reviews, covering a hardware or software design, is not that managers, technical gurus, or other team members pass or fail a design. No, the real benefit of a design review is to let you, the designer, see your design from another, more critical viewpoint and consequentially find critical errors.

In my experience, most errors are found not by the reviewers, but by the designer under review. So don't be afraid of reviews. Actually, they're a powerful tool for you to ensure a correct design.

If your company doesn't have a formal design review process or if the process is too late in your design cycle, try to hold informal design reviews on your work. Do a trade with fellow developers: I'll show you mine if you show me yours. A review will help you to catch early those conceptual or logic errors that can be devastatingly costly to find and fix downstream (an error caught in production costs 10 to 100 times more than fixing it in engineering; the cost rises to 1,000 to 10,000 times for errors fixed in the field).

Documentation Daze

Why did they do this? It doesn't make any sense!
— Typical software user reaction

It's time to re-examine just what hardware/software documentation is and what role it should play. Today we are living with too many old and preconceived notions of documentation that are no longer appropriate for today's larger, more complex products. Think again.

We still do documentation the way we did it in the old days — everything you want to know is crammed into a large pile of manuals. Unfortunately, these days engineers and programmers have more hardware and software than they can cope with. The documentation rules have changed, but someone forgot to tell the hardware and software vendors.

Back in the old days there wasn't a lot of processor hardware, logic, or software. The implicit deal was: if you document it in depth, then I'll read it. Good documentation then meant down-to-the-metal explanations poured into mounds of manuals.

But that deal no longer holds. For one thing, these days, we have a heap of different processors, a river of logic, and uncountable piles of software. There is more hardware and software than anyone can possibly track or use. For another, an increasing number of technical people don't want to read documentation anyway. They want their hands held or expect on-line, interactive documentation to speed up product use. Unfortunately, on-line documentation has a long way to go; most of it is little more than expanded hypertext. (Hypertext is like a dictionary — you need to know the item before you can look it up.)

If you want folks to use your documentation, you had better find a new way to do it — no one has the time to plow though yesteryear's one-size-fits-all documentation.

Don't Care More...

It is much easier to apologize than to ask permission.
— Admiral Grace Murray Hopper, U.S. Navy (deceased)

This is tricky: Care more about your company or product than your management does at your own risk.

When you care more about the company, the technology, or the project than your management does, you have a mismatch, one that can generate serious misunderstandings. You and your management will be talking on two different levels.

If you're any good, you're going to care, so I'm not saying don't. But if you do care more than your management, you're in a situation with a lot of potential hazards; the more you push for success, the more uncomfortable management will be. I've been there too, in situations where the 9-to-5, take-it-easy guys wanted to drift, whereas I wanted us to rock-and-roll, to do aggressive design. It's tough.

1. If your management appears not to care about key areas you're pushing for, you have a mismatch. Such

situations get worse, especially if they are in areas not in your immediate control.

2. If you succeed, management may or may not reward you for it, because they don't see the importance of what you have done. Be patient and go off and tackle another project.

3. If you fail, they'll blame you for wasting resources on something they did not give a high priority to. So pick your targets carefully.

4. Push for those things that you can succeed in. Second, push first in the arenas that are assigned to you; build up a reputation for success, then go after the other areas.

5. Don't, whatever you do, let the do-nothings talk you into drifting with them. In that lotuslike surrender lies long-range, professional suicide.

6. If the situation is intolerable, find a better managed group in your company, or worst case, start looking for a new job. Management that doesn't care about success will eventually become nonmanagement. The only question is what they'll kill on the way down — the group, the division, or the whole shebang?

7. Always consider the possibility that your management may be right. What do they know that you don't? What problems and opportunities do they consider important? Yes, complete idiots sometimes do manage to get promoted to the top, but then so do a lot of competent men and women. Don't be too hasty in your judgments.

8. Your product idea may be right for the marketplace, but wrong for your company. Companies have limited resources and market bandwidth. They may choose to disregard seductive markets and products, so as to keep to their own knitting. That situation — an internally

rejected product idea — has generated innumerable start-ups, as well as whole new industries.

80/20 Rule

Sturgeon's Law: 90% of everything is crud.
— Sci-fi proverb attributed to Theodore Sturgeon

One of nature's little jokes on civilized man. Ignore it at your own risk.

The 80/20 rule says that most activities and actors break down into an 80% – 20% ratio, such as:

1. 80% of the sales are for 20% of the product line.
2. 20% of the sales force does 80% of sales.
3. 80% of code execution time is spent in 20% of the code.
4. 20% of the hardware dissipates 80% of the power.

This is a handy rule to remember; applying it can save you both time and trouble. Called Pareto's Rule (the law of maldistribution), it gives you a valuable rule of thumb that applies to a wide range of activities. Vilfredo Pareto (1848 – 1923) applied mathematics to economic analysis.

80/20 Rule Corollary

A great deal of talent is lost to the world for want of a little courage.
— Sidney Smith (1771 – 1845)

The nail that sticks out is hammered down.
— Japanese proverb

In many cases the 80/20 rule also applies to people and competence: 20% of the people are competent, 80% are marginal. Needless to say, the 80% marginal folks spend most of their time trying to constrain the competent 20%.

Of course, in technical fields such as engineering and software, the percentages of competence are much higher. However, the 80/20 ratio may hold for the ratio of innovative vs. mainline developers. Here, only 20% are innovators, while the majority spends its time trying to rein in this innovative minority.

The tyranny of the 80% is generally not mean spirited; it's quite natural. Innovators tend to upset the status quo and do unsettling things, while the majority values the stability of continuing existing methods and mechanisms. The clash is a natural extension of two differing types.

Today's fast-changing technical world no longer rewards continuity over innovation — both are needed. Even Intel with its highly successful x86 microprocessor line, which dominates personal computers (continuity), had to adapt RISC** technology (innovation) to compete with RISC-based workstations and micros.

If you are an innovator, be aware of the tyranny of the mainstream and recognize it for what it generally is: a queasy discomfort with change, not envy. You can increase your chances of success by playing to their need for continuity, rather than stressing change.

**RISC — Reduced Instruction Set Computer, an emerging processor design technology for creating fast computers. It relies on simplified computer architectures (consequentially less logic and faster cycles) for higher execution speeds.

Engineering
Notebook

By the time the people asking the questions are ready for the answers, the people doing the work have lost track of the questions.
— Norman R. Augustine, Martin Marietta
(Augustine's Law # XXX)

Get an old-fashioned, bound engineering notebook and use it. Carry it everywhere, and use it as a log. Write everything in it from management requests to interchanges with project team managers and members. List all work in progress and any changes needed.

The idea of an engineering notebook is so fundamental that I almost didn't include it here. Keeping a logbook is a sign of a professional. It also helps to minimize misunderstandings and provides a record. Some companies even require technical people to use engineering notebooks and even go so far as to have periodic reviews of them for patent purposes.

A logbook keeps management and fellow coworkers honest. They know that you have a record and act accordingly. It's also a wonderful crutch for a bad memory. Periodically xerox your log and keep a backup record. Needless to say, don't put anything in a log that you wouldn't say to others.

You can use a notebook PC or MAC if you wish (just be sure you back everything up). However, a file-based logbook is just not as believable as a bound log for obvious reasons (such as rewriting history). Notebooks, however, are great for recording, distributing, and filing meeting notes.

Expecting Too Much

The world is a tragedy to those who feel, and a comedy to those who think.
— William Shakespeare (1564 – 1616)

Don't become alienated from your organization because it is imperfect. Just about every technical professional I know, myself included, suffers from unrealistic expectations. Somehow, when it comes to our companies we expect — no, demand — that things work rationally. And when they don't, we go ballistic, as if some key compact with our idealistic soul was violated.

The same folks that take smarmy national politics in stride are outraged when their organizations display similar dysfunction. Somehow we seem to expect, quite unreasonably I might add, our technical world to be different, to be better run than real life, to be closer to technical purity. It isn't.

You programmers know that when you bring up your software, it'll bomb out and twist into untraceable spaghetti; and you hardware jocks expect your breadboards to be brain-dead or smoking on power-up. No surprises there: we all know what happens. And after the fireworks, we simply screw up our resolve and fix what we can.

Take the same attitude toward your working world and its political gyrations. In one word — relax. When things don't work, hey, just shrug it off and do what you can. But when things do work, celebrate — go out and kick a few high ones off the bar.

The very same Murphy's Law that dictates that your hardware or software won't work out of the box also permeates the management hierarchy. Remember, in the long run your hardware and software will work, and so too will your company.

Fear

No passion robs the mind of all its power of acting and reasoning as fear.
   — Edmund Burke (1729 – 1797)

Don't worry, they can kill you but they can't eat you.
   — Anonymous

You can't work effectively under a regime of fear. If you are afraid to take chances, then you can't take the steps needed for a good design or a product to succeed. And so in the long run you'll be ineffective and a prime candidate for layoff.

Managers who run a regime based on fear make a serious mistake. Folks who are afraid will not be effective. And the others, those who are not afraid, will be not be driven by scare talk. Thus, the hard-charging manager is simply wasting his or her time.

Fear doesn't work. That's one of the key lessons from the collapse of the Soviet Union and communism. In this fast reaction, high-tech world, fear slows technical reflexes and erects barriers to excellence. If a company or manager stresses safety over technical excellence they'll get staid, safe, obsolescent products. Fast market reactions and excellence require freedom to act, not constraints that limit action.

If you work for a manager who operates by fear, either set up your own effective relationship or get out. The odds are against your succeeding in that kind of environment; moreover, the organization may not survive either.

Feedback

Feedback is the breakfast of champions.
— Kenneth H. Blanchard, Robert Lorber
(Putting the One Minute Manager to Work)

You can't get better unless you get feedback on the strengths and weaknesses of your work. A control system needs negative feedback to stay on target, so do most technical people. Unfortunately, feedback is harder to come by these days. Honest performance reviews are falling victim to increasing regulation of the workplace.

Many companies now routinely use layoffs or reorganizations as a stealth way to get rid of perceived nonperformers. These so-called nonperformers are being shortchanged: they get no feedback to correct their deficiencies. Unfortunately, an increasing casualty is in-depth performance reviews.

Engineering and programming, are still performance-driven functions — if the work isn't done, it's pretty obvious. Even so, managerial feedback will probably slack off a part of the general trend. The best way to protect yourself, to get better, is to seek feedback on your performance from your peers and managers. Don't be afraid to find out what you did right, and conversely, what you did wrong. You can't better your skills unless you know what to improve. So ask for feedback. Be persistent in seeking it out. You need it.

Find a Good Critic

A guest sees more in an hour than the host in a year.
— Polish proverb

Friends are a treasure, but a good critic is worth his or her weight in gold.

Before you submit a design or a proposal, find a critic to go over it and highlight potential flaws. Needless to say, it's far better to find errors early than to live with their effects downstream.

Nobody is perfect; not you, not me. The secret to workable perfection is a good critic to head off the mistakes or lapses in judgment that we're all prone to do.

Flattening

Man is a complicating animal. He only simplifies under pressure.
— Robert Townsend, AVIS

American (and European) companies are busily flattening their organizations. Too many layers of management between top and bottom had lead to sluggish, slow-to-react companies that were unable to compete in the fast-moving 90's. Today, companies are on a diet, and are squeezing out those excess management levels.

At the same time, automation, more powerful computer systems, and better design/management/support tools enable fewer employees to do far more work. Today, it takes fewer and fewer people to do design and implementation. Consequentially, companies are also slimming down their technical and production teams.

Similarly, computers, communications technology, and software have combined to eliminate many middle management functions. Systems built on these technologies now routinely collect, manipulate, and dispatch monitoring results to higher ups. These were tasks routinely handled by middle management and staff.

Lean and mean will be the norm for successful organizations. Wal-Mart is a good example; it's run from a central control room that looks like the tower of a major airport. There nationwide store transaction data is gathered and used to refill shelves and control shipping, as well as to set promotion strategies. Wal-Mart's top management has the data and mechanisms to run large numbers of stores without the traditional white-collar staff and management layers.

If you're a technical developer, don't expect to easily move into management roles — there are fewer management slots. And don't expect to loaf along at work either. Flattened, downsized organizations run lean and mean. If you like to be able to get things done, if you dislike constrained organizations, then you're going to have a good time.**

** For more on flattening read *Rebirth of the Corporation* by D. Quinn Mills or *The Virtual Corporation* by William H. Davidow & Michael S. Malone (see Bibliography).

Fun

Work is much more fun than fun.
— Noel Coward

We're going to have serious fun.
— Gene Wang, Symantec

If you're not having fun doing what you do, find something else to tackle.

Technology — designing it, implementing it, keeping it running, and managing it — can be a lot of fun. It keeps changing, inexorably moving up a performance and experience curve. That kind of challenge keeps work from being stale and fixed.

If you're bored, if you can't keep your eyes open, if you live for weekends, then save yourself a lot of trouble and find something else to do. Find a challenge that gets your juices flowing. If you hang around and go through the motions, you'll eventually get hung out to dry.

Developers at Borland, Intl. (Turbo C/C++, Turbo Pascal) came up with what I think is the perfect definition — serious fun. Translation: this is serious business, but dammit, let's have fun doing it.

Gearing Up

*The constant temptation of every organization is safe
mediocrity. The first requirement of organizational
health is high demand for performance.*
— Peter Drucker, Management guru

This technical world is not a democracy. The most successful
companies are on their way to becoming meritocracies —
places where the best people are given the leeway to produce. As company hierarchies flatten and organizations strip
down to fighting weight, they are also gearing up to make
use of their best people. In the past, technical companies followed the military lead and built organizations that empow-
ered average managers and engineers to turn out successful
technical products.

Unfortunately, we no longer can afford the drag of mediocre management and indifferent technical work. We no
longer can tolerate the built-in fat and inertia that enables
average management, run-of-the-mill engineers, and mid-
dling programmers to succeed. Relying on the mediocre has
led to less than stellar products, plodding product development, and stifling environments.

It's not unusual to have production disparities of 10 to 1, even 100 to 1 between individual programmers or between individual engineers. Thus, it only makes sense to try and take advantage of these high productivity people and not slave them to lesser standards.

We need to build the best products we can, not just enable run-of-the mill managers and technical grunts to succeed. The movement is already under way to shift from standard infantrylike organizations to empowering more elite teams — as the Army has done with its Special Forces and Ranger teams.

Ground Floor Opportunities

True artists ship.
    Steve Jobs, cofounder, Apple

The good news about start-ups is that you can have the experience of a lifetime, and if lucky, can make some significant bucks. The bad news is that start-ups can easily drive you around the bend.

Join a start-up if you:

- Have an excess of energy.
- Are bored stiff (if you spend Friday and Saturday nights at home watching TV, you're a good candidate).
- Are really good at what you do and want a chance to really show it.
- Want a challenge that will totally dominate your time and life.
- Want a chance to make it big dollarwise.
- Can work 48 hours straight, kept awake by bad coffee or Jolt Cola.

Conversely, keep away from a start-up if you:

- Are an 8 to 5 person.

- Want a significant home life.
- Can't live with crazies.
- Can't take merciless live-or-die schedule pressures.
- Can't take chaotic management and market pressures.
- Can't live on pizza, Chinese food, Jolt Cola, and candy bars.

You can make a lot of money at start-ups or companies that are just ramping up. A number of programmers at Microsoft and engineers at Apple became millionaires from their stock options. Even in today's more settled times, technical folk can make big bucks at start-ups — that is, if they are willing to pay the price.

Start-ups are generally started by corporate barbarians, raiders who see a technical opportunity to pillage and burn. Gifted and driven, they're not what you would call solid managers. In start-ups you run with the pack, put your shoulders to the product wheel and move product. There's nothing more professionally gratifying than to be part of a start-up that succeeds — it's an experience of a lifetime.

But start-ups tend toward the chaotic. The more advanced the technology, the wilder the ride. Most start-ups focus on product and technology, not people. You have to fend for yourself and be almost self-managing to succeed at most start-ups.

Finally, being a plank owner (an early employee) doesn't guarantee you longevity or even a major say at the resultant company. Start-ups generally evolve from quick reaction, technology-driven management to more staid, corporate management (known as the "suits").

Hairy Arm

Sometimes you have to give some to get some.
— Business proverb

A good proposal may not be enough to succeed.

There's an old apocryphal story about an Italian Renaissance painter who made his living painting portraits. Four times he painted the portrait of his patron's family, and each time the patron rejected it. Finally, on the fifth try, the desperate artist added a hairy ape's arm firmly around the wife's portly figure. On seeing the portrait, the patron stared and stared and stared. Finally, he exclaimed: "That portrait is good, but it has an incredible flaw," and pointing to the hairy ape's arm, he said, "Fix that and I'll take it."

For some folks to buy into a proposal, to accept something, they need to put their own mark on it. Thus, it may pay to leave one or more obvious flaws for them to fix, and thus make an investment in your project. A perfect work defeats them because there's nothing left for them to do but criticize it.

Hiring

*People get hired on the basis of skills and get fired
on the basis of personality.*
— Common industry knowledge

*If each of us (other executives) hires people smaller
than we are, we shall become a company of dwarfs.*
— David Ogilvy, Ogilvy & Mather

Some advice for hiring managers:

1. Hire for personality and for drive first, for technical
   skills second. A good developer can pick up the extra
   technical skills as needed, but people almost never
   change personalities. I know managers who hired in
   critical technical skills and ended up with a problem
   child or a festering sore. If you absolutely must have
   a critical skill in-house, consider hiring a consultant
   and using him or her to train a current employee.

2. Look for points of craziness. Just about everyone has
   some points of discontinuity, nodes of irrational
   behavior. The interviewer's task is to identify these
   points and determine if they are debilitating or not.
   It's pretty easy to find them. Here's a technique
   called trolling. For example, for problems relating to
   management, just say: "All managers aren't good,

sometimes they can create some real problems..."(you can give an example of your own experience). And then just sit back and listen to what they say. You can apply trolling to any key areas such as working with others, debugging, office policies, etc. It's surprising how well it works.

3. Headhunters can save you time, but you must make them work at it. Many headhunters take the lazy way out: they just throw resumes against your wall to see if any will stick and lead to a hire. You could do the same thing with ads and a hired temp.

   On the other hand, if headhunters do initial candidate screening, they can save you a lot of time. Give them detailed specs and demand that they screen applicants first. Put it on a three strikes and you're out basis. If they can't get into the ballpark with the first three applicants they send in, they're out.

4. One of the most valuable services that headhunters provide, is to keep potential hiring situations alive. They maintain a live link between you and the applicant, and iron out any misunderstandings that could block a successful hire. A good headhunter interviews applicants after you do and provides you with detailed feedback on the applicant's reactions and plans.

5. In interviewing candidates, look for curiosity and increasing technical skill levels. The skills in demand today may not last for tomorrow. Long-term employees should be able to grow in skills and have the ability to tackle a range of problems. The president of a small CASE (computer aided software engineering) tool house routinely asks applicants what technical books, journals, and magazines they read, as well as any seminars or classes they take. If they

don't show much external technical interest and activity, the interview is over, they've struck out.

6. Look for people who are doers: engineers or programmers who take responsibility and finish projects. Beware of applicants who talk of helping or aiding projects. Ask for hard figures that size their projects. And ask for peer as well as for management references (peers may be more willing to talk than managers are about a person's technical and interpersonal skills).

7. If you build commercial products don't automatically discount aerospace managers, engineers, and programmers. There are some very good, tough-minded aerospace technical folks out there. Look for the rebels and the doers who succeeded in spite of the glacial aerospace environment. Think of them as training in a high-gravity (resistance) environment; the technical and managerial muscles they built up could be pretty powerful in lower gravity commercial environments.

I Don't Know

Looking dumb is oodles better than lying.
— Rear Admiral Dave Oliver, Jr., U.S. Navy

"I don't know." Three of the hardest words in the English language for many of us, especially technical people, to say.

If you don't know, say so. Don't try to fake out people, especially your managers. They are not stupid: they cannot expect you to know everything. However, don't get caught short twice on the same topic — that can downgrade you.

Whenever you get a question that you can't answer, just say: I'm afraid I don't know, but I'll find out. And then do so.

You can't be perfect — no one can. But you can be effective. Making mistakes or not knowing everything is small potatoes if you get the job done right — that's what really matters.

If It Ain't Broke...

When you see a snake, kill it.
— H. Ross Perot

There is no right way to do the wrong thing.
— Business proverb

If it ain't broke don't fix it. Wrong. Improve it, get better.

If it ain't broke, don't fix it was the battle cry of American industry at its heyday, in the halcyon 1940s and 1950s. These days, it's passé or worse — the slogan of companies wallowing in economic doldrums.

Relying on periodic large-scale changes or reorganizations for improvement has become the modus operandi of most American companies. Why? Because management hierarchies have become rigid and bloated with numerous management levels (see Flattening). These rigid management structures are incapable of dealing with day-to-day change or adapting to rapidly changing market conditions.

In contrast, many Japanese companies have learned to use continuous, incremental change. When they find a minor improvement or a small flaw, they fix it; that tactic has lead to world-class manufacturing and product technology.

Incremental change is a good habit to have. It's easy: when you find a problem,** fix it. Take out the minor glitches, and a lot of towering barriers become scalable. You have to start somewhere.

** Incremental change, fixing low-level problems, may not have a major effect if the problems are really symptoms that mask high-level, structural, or process problems. For example, there may be no easy fixes to enable a rigid, hierarchical organization to become a fast-reaction, market-oriented competitor.

Imprinting

Arabol que nace torcido, nunca su rama endereza.
(A tree that grows crooked never can straighten its limb.)
— Mexican proverb

If the blind lead the blind, both shall fall into the ditch.
— Matthew 15: 14

Ducks do it, so do humans. Newborn ducklings imprint on the mother duck, following her and absorbing critical survival skills. Similarly, newly hatched engineers, programmers, and other technical folk imprint on their managers, taking management and organization cues from them. Therefore, a few bad, first-level managers can literally poison a generation of technical talent in a company.

An incompetent or mad dog first-level manager can inculcate bad management practices in his or her charges. Later, as these people move up to management, many will promulgate the same bad management practices that they grew under. Just as dysfunctional families can lead to next-generation dysfunctional families, so can dysfunctional managers create future dysfunctional managers. A company's technical talent represents its real capital. Letting a bad manager waste that capital is almost suicidal, as well as just plain bad business.

If you're starting out and are stuck with a bad manager, don't let him or her sour you: look around for a better managed group and move there; talk to your manager's manager; read a good management text** for a balanced view.

** I recommend: *The Effective Executive* by Peter Drucker, 1967, Harper & Rowe, NY; *In Search of Excellence*, by Thomas J. Peters & Robert H. Waterman, Jr., 1982, Harper & Rowe, NY; and *Further Up the Organization*, by Robert Townsend, 1984, Alfred A. Knopf, NY.

Indirection

Throughout the ages, effective results in war have rarely been attained unless the approach has had such indirectness as to ensure the opponent's unreadiness to meet it. The indirectness has usually been physical and always psychological. In strategy, the longest way round is often the shortest way home.
— B.H. Liddell Hart
   (Strategy)

Many think that military strategy has a direct application for commercial marketing and the clash of competing companies and products.

For an intuitive view on military tactics and history, read the classic work on military history and strategy: *Strategy* by B. H. Liddell Hart. Liddell Hart was the military editor of the *London Times*; Along with Charles De Gaulle of France, he is credited as being one of the originators of Blitzkrieg or tank warfare that the Germans used to win the initial stages of WW II (the Germans had paid attention, but the French and British hadn't).

Strategy covers military history from the early Greek wars through the 1957 Arab-Israeli war. Liddell Hart's thesis is that direct assault, direct attack almost never wins. And that all through history, indirection has been the winning strategy.

In business, head-to-head competition may succeed, but it is almost always costly. Indirection -- changing the shape or nature of the competitive battle -- offers a more cost-effective way to victory. For example, new industries and technologies generally offer a new and different benefit set than the existing vendors. They compete on a different battlefield, and as they succeed, they eventually obsolete and bypass older products.

It's considered very difficult to sell directly against an entrenched product, one with a sizable, established market share (much like assaulting a fort). Basically you end up fighting on the competition's choice of ground and will probably bloody yourself for marginal gains. Strategy says use indirection; compete on grounds more advantageous to your product.

In day-to-day technical work, indirection may be the best way to win the internal conceptual or methodology wars. Instead of arguing about which concept or method is superior, maybe a better way is to show its strength in other, supporting areas, to flank the argument.

Strategy is still in print; a quality paperback was reissued during the Gulf War (the US and its allies used indirection to win).

Innovation

Imagination is more important than knowledge.
— Albert Einstein

An ounce of innovation is worth a ton of code.
— Kamran Parsaye, Intelligenceware

These days engineers and programmers seem to be under the product turnaround gun — there is little time for significant innovation. Today's product cycles, both hardware and software, are shrinking, and folks have to turn to and hustle product out the door. Even worse, many individuals and companies seem to be risk adverse, afraid to try new approaches.

Another consequence of compressed product schedules is a narrower focus for most technical folk. They have less time to read widely, to explore other, neighboring technologies. And that's a shame, for many breakthroughs and innovations combine disparate technologies and methods.

In his classic book *Creativity*, Arthur Koestler defined creativity as bi-association — the association of previously separate ideas, technologies, or mechanisms. Examples of bi-association include: Gutenberg's invention of printing with movable type; Kepler's synthesis of astronomy and physics;

and Darwin's theory of evolution by natural selection. (All innovation, however, is not the result of bi-association; key innovations also emerge as developers acquire deeper and more intimate understanding of the problem and its constituent relations.)

As I write, this I've been a technical editor covering electronics for over 9 years. News and details of most new technologies and products pile my desk. And I've noticed a lack of major technical innovation — there is seemingly not a lot of new, innovative products or technologies. Most products that cross my desk are incremental advances, iterations of current technology and products.

This dearth of major innovation presents an opportunity for those willing to push product limits. Keep your curiosity about other areas; read in them. Who knows — you may find a new innovation, a symbiosis that delivers large product strides.

Integrity

Character is power.
— Booker T. Washington (1856 – 1915)

To thine own self be true,
And it must follow, as the night the day,
Thou canst not then be false to any man.
— William Shakespeare
(*Romeo and Juliet, Act II*)

Without integrity you can't go far in this technical world. The most important thing you possess is not your technical skill (that's second), it's your integrity, a basic honesty and forthrightness that people can rely on.

So don't compromise it. Tell the truth (see Nothing but the Truth), don't hurt people and keep your word. Live by high standards and you won't regret it. As businesses go, this technology world is a pretty honest place. Honesty is the rule, not the exception; we have to trust one another to be able to pull products together (a major reason why corrupt societies such as Russia have difficulty in fielding competitive, high-tech products).

Remember, no matter how difficult today is, there will be a tomorrow. But if you compromise yourself for today, you may find yourself crippled for tomorrow. For example, during layoffs I've heard people say that they'd do anything, hurt anybody, do whatever it took to keep their jobs. Well, they kept their jobs, but were never trusted after that. This is really a very small world of companies interlocked by job trails and associations. Never assume that what you do at company X won't be known at company Y. In fact, you'd better count on it being known.

People who steal or cheat on minor matters are horrified to find themselves ostracized and not trusted on major matters. It just never occurred to them that the two were linked. They are.

Dishonesty is dishonesty. It's like being a little bit pregnant or the old joke of the philosopher and the movie star:

> Would you go to bed with me for $2 million dollars?" asks the philosopher.
>
> "Sure," says she.
>
> "Well then how about for $5?"
>
> "What do you think I am?" she snaps.
>
> "We've already established that," explains the philosopher, "now we're just negotiating price."

Invest 30 Minutes
a Day

Knowledge keeps no better than fish.
— Alfred North Whitehead (1861 – 1947)

If you think education is expensive, try ignorance.
— Derek Bok, Harvard

It's easy to get caught up in the daily struggle, to focus on chopping your way through the design jungle, and to making way for your design. Time flies; months pass; the seasons roll on by. When finished you have the knowledge and strengths built up during the project, but you could have more: a wider knowledge picked up by investing a small amount of time each day.

Pace yourself. Make room for a half-hour, an hour a day. Set time aside to keep technically current, to read your technical journals, and to explore other technologies. Do it. If you're in deep yogurt on your project, a half-hour won't save you. In fact, it may relax you enough to re-enter the fray refreshed with a mind cleared for battle. Wait until you can't look at your code or logic any more, then take a technology break.

Exploring other technical areas or monitoring the technical press is not a waste of time. It may give you an added edge for future work and technical directions. One facet of innovation lies in the integration of dissimilar technologies for a new, more powerful synthesis. Where do you find new candidate technologies for integration if your focus is on a narrow specialty? You can't (see Innovation).

Moreover, today's wide, uncharted specialty has a way of becoming tomorrow's narrow known base. Technologies are explored, tamed, and turned into common knowledge. Stay ahead of the curve — there's opportunity out there.

Irreplaceable Man

There is no indispensable man.
— Woodrow Wilson

The graveyards are full of indispensable men.
— Charles De Gaulle

If you're irreplaceable at work, you may already be on your way out the door.

If you're invaluable for an advanced skill, or for the capability to define new product technology or for your insight — then you're fine. But, if you're counting on permanent employment because you're the keeper of the rules for some arcane software tool or for your knowledge of past-done work or procedures, think again.

Design technology is moving fast; yesterday's knowledge is already on its way to the technology rubbish heap. If you want to survive and thrive you must continue to grow, to add new skills and capabilities to your design quiver.

You can't do that sitting around as the keeper of the keys. You simply get more expensive as your salary goes up each year. And at some point management may wonder why they need you around at all, especially when they finally junk the system or methodology that you've been slaved to for all those years.

It's Not My Job

Your job is to get the job done.
— Silicon Valley proverb

The dodo and the passenger pigeon are extinct; going next
are those work places and employees that substitute rules
and boundaries for thought and collective effort. If you want
a slot on the "deadwood" list, just parrot "it's not my job"
when things go bang in the night as projects move down the
product home stretch.

Defending your job and organizational boundaries made
a sort of convoluted sense in days past when companies were
Balkanized into small fiefdoms that made up a sequential
product development process. These days, organizations find
that they can't afford the overhead and inefficiencies of such
highly compartmentalized and rigid structures. Instead,
companies are evolving toward product/project or market-
segment organizations or clusters (see Clusters).

Today, the trick is to find the balance, the right mix
between increasing your job efficiency and company product/

111

service throughput. In the olden days, it was simple: you could defend your department and job boundaries in the name of increasing your professional effectiveness. No more. After all, what good is your or your group's effectiveness if the company hits the Chapter 11 wall?

Obviously it makes sense to sacrifice local efficiencies to ensure company survival and market presence. On the other hand, you can overextend yourself in doing whatever is needed, and not be cost effective or grow technically. You sure don't want to end up as the most expensive carton stuffer in the building.

It's Not What You Do ...

Here lies a man who knew how to enlist in his service better men than himself.
— Andrew Carnegie's tombstone (1835 – 1919)

It's not what you do that's important if you're a manager; it's what your people do.

That concept is a hard lesson for new and even in-place managers. It's funny really: we generally promote the best, i.e., the most effective technical people to positions of management and then expect them to somehow get others to do as well.

Many times this variation of "set a thief to catch a thief" (set a productive person to manage others to be productive) does work. Sometimes it does not. The problem is that many personally effective people have trouble letting go, letting subordinates make mistakes and do the best that they can.

These managers unconsciously use themselves as the standard and assume that others can perform at their, the manager's, previous level. They don't realize that they are paid not for what they personally can do, but rather for what their people can do. If they're not careful they can become obsessively detail- and control-oriented in trying to make their subordinates work at the manager's level. This over-control can block subordinates from developing into effective workers, as well as turn into a dandy roadblock.

Kissing Up

*Never pick a man because he slobbers all over you with
kind words. Too many commanders pick dummies for
their staff. These dummies don't know how to do anything except say "yes." Such men are not leaders. And
any man who picks a dummy cannot be a leader. Pick
the man who can get the job done.*
— General George S. Patton, U.S. Army **

Everybody hates a brown-noser, especially in a technical
world, which by definition places such emphasis on results.
It's not necessary to kiss up to get promoted or to get a hearing for your ideas. Actually, many technical people hold their
management at a distance because they don't want to be perceived as kissing up. That's foolish; it doesn't give managers
a fair chance at successful management. Here's what is and
is not kissing up:

It is not brown-nosing to:

- be polite to your manager and make him or her comfortable.
- avoid continuous confrontations with management.
- carefully pick the arenas to differ with your manager to
  change his or her policies.
- go out of your way to make contact with your managers and
  let them know what you are doing.

- help your boss to be successful.
- congratulate your boss on something she or he did well (they get few honest compliments; they need feedback too).

It is brown-nosing to:

- be obsequious.
- be a yes-man (or woman) and tell your boss what he or she wants to hear rather than what they need to hear.
- avoid differences of opinion on important items.
- side with your manager against fellow subordinates when the situation doesn't justify it.
- focus on relationships with your managers rather than on the results of your work.
- avoid bringing to your management troublesome situations that need immediate attention.

If you spend all your time doing the political thing rather than the effective thing, the time will come when you must act and you won't know how.

** If you want to get a feel for Patton's management style read: *Patton's Principles: A Handbook for Managers Who Mean It!* by Porter B. Williamson, 1979. Simon & Schuster, NY.

Know When to Stop

The road to Hell is paved with good intentions.
— Samuel Johnson (1709 – 1784)

One of the hardest things for most of us is to admit when we're in over our heads, to recognize a bad technical start or direction, or to acknowledge a bad product idea. As Molly Ivins, a popular columnist noted: "The first rule of holes: when you're in one, stop digging."

Learn to monitor your progress, as well as that of a project or technical direction. Remember, it's better to stop a bad project and redirect resources than to eventually suffer a product failure.

Good managers know that they are really paid not to closely direct their troops, but rather to step in and help individuals or projects, when they slide into the morass and start floundering.

Good senior engineers or programmers know when to stop, when to fold their cards and call for a new project deal— that comes from experience (read failures). But, juniors and intermediates may not. Consequentially, they need a closer watch to ensure they don't take a wrong turn and end up too deep into the design swamp to back up and try again.

So back off now and then and try to get an overview of how well your project and individual work is going. Don't be afraid to ask for help or a design review. Don't be afraid to retrench and back up for a better solution. Remember, it is the final result, the product, that counts, not how you got there.

Lateral Thinking

It is an old maxim of mine that when you have excluded the impossible, whatever remains, however improbable, must be the truth.
— Sherlock Holmes
  Sir Arthur Conan Doyle (1859 – 1930)
  (*The Adventure of the Beryl Coronet*)

Think sideways!
— Edward de Bono

There you are: it's 3 A.M.; your mind is slowly turning to mush, and your program or your logic doesn't work. Again and again you try it — A, then B, then C, then ...— and over, and over, and over. Not only does it not work, but the sequence of program or logic events (A -> B -> C -> G) has become so deeply grooved into your brain by constant repetition, that you couldn't recognize a solution if you fell over it.

It's easy to paint yourself into a solution corner: to be trapped, endlessly trying and retrying solutions that don't work. One way out is to go at your problem in different ways or directions, to try a bottom-up or an inside-out approach.

Lateral thinking says: stop — break that causal chain, go at the problem from another angle or a different tack. Break up that pattern; suspend judgment; free up your mind and try combinations outside of the fixed sequence. Go backwards: go to G and ask why did C get here. Or try changing initial conditions as there may be some hidden assumptions there; go in the middle and see what happens; or even go so far as to set up illegal conditions and see what pops.

Lateral thinking encompasses a set of concepts, a way of solving problems, a pattern-breaking way of thinking. Pioneered by Edward de Bono, it's now taught in some schools as well as at major corporations. De Bono has written a series of books on lateral thinking over the years. His best book is the first: *New Think — the Use of Lateral Thinking in the Generation of New Ideas,* Basic Books, 1967. A short, easy-to-read book, you can read it in a few hours and walk away with a different approach to problem solving.

So if you find yourself facing a seemingly unsolvable technical problem, try taking a different tack. Instead of following sequential thinking (A implies B, B implies C, and C implies G), try and break the chain. Go from results to first principals, get a little crazy, it might break the conceptual logjam. Shake things up a bit: Toss out or permute assumptions and see what happens.

Lazy Does It

The Philosophically Lazy shall inherit the earth.
— The author

Laziness, rather than necessity, is often the mother of invention.
— Robert Noyce, Intel cofounder

High-tech endeavors need the lazy for continuous improvement. There is nothing worse than folks who will take any system and make it work as is. Like army ants, they'll swarm all over and make it work by sheer numbers and effort. They'll get it to work, but nothing will ever change, nothing will ever get better.

Needed are what I call the Philosophically Lazy — the high-energy technical folks who have a lazy bent. These are the people who work like mad, but always are on the hunt for easier ways to do things. Instead of doing things in the standard, plodding rote way, they will bust their tails to find a shortcut, one that satisfies their technical sensibilities.

Take a look around your company; you'll find a variety of workarounds, shortcuts, and special tools. Guess who did them? The odds are that they were implemented by the Philosophically Lazy.

The Philosophically Lazy are not exactly lazy. They probably put in more time to do things (at least initially) than the army ant type folks, but they leave a legacy of shortcuts. With the straight-arrow types things get done, but they never get better — the system never changes.

So value your lazy folks, they may be the best resource you have.

Learn to Write

It's O.K. to be a bad writer if you have something to say.
— Bill Stott

All is fair in love, war and getting started writing.
— Gene Olson

Whatever we conceive well, we express clearly.
— Boileau

1. Most engineers and technical people can't write for beans. Many look down on good writing as someone else's problem. But no matter how good you are technically, no matter how articulate you are, it will be much easier to get your way if you can write well.

2. Remember, you're competing against a host of others, many with similar technical skills and drive. If you can write well, then you have an additional skill, one that easily distinguishes you from the pack.

3. Writing just ain't that hard. There are on-line text editors to fix your grammar and spelling. Thinking — that's the hard part. Take care of the thinking side and you'll be surprised at how easy writing becomes.

4. Writing can be learned. Raymond Chandler, the dean of the hard-boiled detective novel, was an outstanding writer. At the age of 44, after losing a battle with

the bottle as well as his job as an oil company executive, he decided to become a mystery writer. How did
he do it? He learned his new craft by simply taking
mystery stories out of magazines, and then rewriting
and rewriting them until his versions bettered the
originals. It worked. To this day, no one has topped
(in my opinion anyway) Chandler's prose or style in
mystery fiction. Others have also used this method:
Benjamin Franklin learned to write well by copying
articles from *The Spectator* into verse and then
translating it back into prose!

As an engineer and manager, I had written only
specs and proposals. Starting out as an editor/writer,
I had two major problems: I couldn't write and I
couldn't type! In a panic, I followed Chandler's lead,
and started out by typing in two newspaper editori-
als each day and then rewriting them until my versions were sound. It worked. As an added benefit, I
also learned to type (sort of)!

5. For those of you picking up writing skills late in life
as I did, there are two major learning stages. First,
you have to aim for simplicity and the ability to write
clean, direct prose. Once at that level, you need to
tackle the next stage and learn how to fold in complexity and art (I know, I know — I'm working on it).

For the first stage, trying to write as we talk is sound
advice. Read *Style* by Strunk & White, and *How to
Write Like a Pro*. (see Bibliography). Both will help
you get started. Needless to say, the second stage is
much harder; for hints, read: *The Art Of Persuasion*
and *Style*.

6. In technical writing, content is far more important
than style or beautiful prose. Many of us were
hounded by a series of pedantic English teachers and
demanding professors who stressed technical mas-

tery over understanding. All well and good; however, if you have something to say, and can organize it, that's over 50% of the writing battle.

7. One of the biggest pitfalls in writing is that we don't see what we HAVE written. Instead we see what we thought we MEANT to write. For that reason, it's best to wait a day or so and review your text before sending it on. One way to check your writing is to read it aloud. Awkward phrasing, inappropriate punctuation, and clumsy structures will stand out, as Chandler phrased it, "like a tarantula on a slice of angel food cake."

8. Finally, to be a good writer you must read; there is no other solution. Without reading others and learning from their work you won't get far. But, reading biography, business texts, and novels can't hurt you. In this life you have two choices: you can learn from experience, which generally means making mistakes; or you can learn from others' experiences and mistakes. Reading gives you access to over 2,500 years of human experience to draw upon. Why not take advantage of it?

Line Versus Service

Because that's where the money is.
— Willie Sutton, on why he robbed banks

Work where the money is: line organizations. They are the segments that directly add to a company's revenue base, and consequentially get management's attention, as well as resources. Service organizations, on the other hand, serve these line organizations and typically get the leavings.

Common sense says to work where the revenue action and management attention is; that is also where promotions and raises tend to be found.

Generally, line organizations are listened to, while service organizations are tolerated; line organizations get needed resources, whereas service organizations usually make do with less. However, this state of affairs is changing as a number of corporations integrate service functions, such as Customer Service, with line organizations. Some flattened organizations take on forms that integrate both the service and line functions into customer-oriented clusters.

Use this rule of thumb: if a group adds directly to the financial bottom line, it's a line organization; if not, it's a service organization.** In technical organizations, line functions include: design, development, product management, product marketing, marketing, and sales. With the current emphasis on product reliability and customer service, product test, application engineers, and field service are considered line organizations by many companies.

** There are exceptions. In today's era of potential catastrophic regulatory effects on the bottom line, many service organizations that deal with regulatory problems are treated as line organizations.

Lists: Stay Off the Deadwood List

I've got a little list — I've got a little list
Of society offenders who might be well underground,
And who will never be missed — who would never be missed.
 — Gilbert and Sullivan
   (*The Mikado* opera)

Stay off the bad guys list. Every company, every organization, every manager has their own little private deadwood
list, a roll call of those to go next. Don't get on it.

Don't be fooled into thinking your management doesn't
notice what's going on. Don't assume you're getting away
with something just because they don't say anything; that
doesn't mean that they haven't tumbled on to it.

Managers don't like hassles. Many simply avoid sticky
situations that involve disciplining people, especially if they
plan to get rid of them downstream. Remember, managers
can bide their time. Eventually those subordinates that find
their way onto the "bad guys" list will pay a price for their
perceived errant behavior.

It doesn't take a lot of imagination to list obvious sins
that qualify one for the deadwood list. These include: not
doing your work, being continually tardy, consistently slip-

ping schedules, constantly causing trouble, a bad attitude, a lack of energy and drive, refusal to share knowledge, not working an honest 40-hour week, an unwillingness to grow, and plateauing skills.

Talking of such lists reminds me of a friend. We both worked for a small system software house where he was the Unix system programmer. He had the innate curiosity of a cat: he'd put his nose into everything. One day while playing around, he broke into some protected files, which turned out to be the minutes of a board of director's meeting. And buried in those files was a list of names, one of which was his. It drove him crazy; he just couldn't figure out what the list was for. A week later he was laid off.

Be good.

Luck

Everything comes to he who hustles while he waits.
— Thomas Alva Edison

*The luck of having talent is not enough; one must
also have a talent for luck.*
— Hector Berlioz

Some people are lucky; some are not. But luck involves more
than just encountering favorable circumstances: it means
being willing and able to take advantage of them.

It also means knowing when not to push your luck,
knowing when it's prudent to withdraw, and not attack.
Julius Caesar was considered lucky. One reason is that he
knew when to press his advantage and when to sit in his
tent. (See *Caesar's Gallic Wars*; it's in paperback and librar-
ies.)

There is an old, apocryphal story about Napoleon
Bonaparte. It seems that he was casting about for a new general to lead one of his armies. His courtiers proposed one
general after another. For general X, they said: "He went to
Saint Cyr (the military academy) and he fought in this battle... and he won in that battle."

131

"Yes, yes, yes," replied Napoleon, "But what I really want to know is: is he lucky?"

Lying to Yourself

Nothing great was ever achieved without enthusiasm.
— Ralph Waldo Emerson (1803 – 1882)

Believe it or not, this technical world is an extremely honest place, especially when compared to many other industrial and business segments. People do lie, but they actually lie more to themselves than to others.

Think of what it takes to bring out a new product or to start up a company. If you really knew at the outset what problems you would run into, you'd probably not even get up out of bed in the morning, let alone go in and do heavy technical lifting. To succeed you have to be optimistic, to look on the positive side, and that's where things can get, as they say, "a little creative."

As a technical editor, I talk to many companies. Yet, I can count on the fingers of one hand the times that I've been directly lied to. However, I wish I had a quarter for every product, software and hardware, that slipped and didn't meet its shipment date — I'd be rich. In general, most of the

133

slipped product vendors had taken the positive view, the more to convince themselves than others.

To succeed you have to strike a realistic compromise. Yes, you need to be hopeful and enthusiastic. But no, you'd best not be too optimistic about schedules. You're caught between two dangers, the Charybdis of unchecked optimism and the Scylla of hard reality.

Make a Friend

Whom they have injured they hate.
— Lucius Anneaus Seneca (8 B.C. – A.D. 65)

No good deed goes unpunished.
— Oscar Wilde (1856 – 1900)

If you want to make a friend, let someone help you; if you want to make an enemy, let someone wrong you without redressing it.

Most of us are in such a hurry to get our designs moving and debugged, that we have little time or patience for people. We pay attention to the tasks at hand, but not to professional relationships. But good professional relationships lead to good products.

Never turn away well-intentioned help; let them make an investment in helping you (you can minimize the time involved). It pays off: they may find something you forgot, add to your design or project, and they will have made an investment in your success.

Alternately, an unpaid debt or the sour taste of wronging someone can lead to actual enmity. People who injure or wrong you may then find themselves thinking of you as an

135

enemy and act accordingly. (This is seemingly odd behavior, as you are the wronged party, but that is how human nature often works.)

So if you find yourself in that situation, try to head off the consequences with simple actions. Let people repay you with help. If someone has wronged you, give them an opening to make amends and realize that you harbor no grudges.

Mistakes

The only man who never makes a mistake is the man who never does anything.
— Theodore Roosevelt

The greatest general is he who makes the fewest mistakes.
— Napoleon Bonaparte (1769 – 1821)

Everybody — you, me, the engineer or programmer in the next cube, the president of your company — all make mistakes. The only difference may be in how we handle our errors once we make them. Hiding them, or stonewalling (as Nixon did) can lead to disaster.

The best thing to do with a mistake is to admit it; fix whatever can be fixed; and then go on with what has to be done. Effective people make mistakes — that's the price of doing things. On the other hand, ineffective people don't make many, because they carefully limit their actions. You can't hit a home run unless you go up to bat and swing at the ball.

One way to limit mistakes is to have your work reviewed, either in design reviews or by your own private critic (see item: Find a Critic).

The more people in place to catch and correct errors, the better off you are. Encourage subordinates, colleagues, and administrative people to find and fix any mistakes you make. People who are aggressive and brook no interference commonly end up in grief, because others simply stand by and let their "don't mess with me" contemporary's mistakes bear its own ill fruit. They won't go out of their way to hurt these hard-to-work-with folks — it's not necessary — they just won't try to save them from the consequences of their own actions.

Nobody is good enough to not need help. Not only is it efficient for others to help correct one's errors, but it also adds to the overall bottom line — it's good business.

Nice Guys

To be humble to superiors is a duty, to equals courtesy,
to inferiors nobleness.
— Benjamin Franklin (1706 – 1790)

Working for a nice guy or lady is a lot easier than working for an SOB. It's certainly more pleasant.

But nice guys and ladies tend to avoid confrontations, or dealing with sticky situations. Unlike the SOB, who revels in letting you know exactly where you stand, nice managers tend to let problems slide.

So if you're lucky enough to work for a nice guy or lady, there are a few things that need tending to. One, make sure you know how they feel about critical issues and, two, follow up on any unresolved issues that could potentially cause problems for you later.

Make sure that your manager knows what you're doing and that you get feedback on your progress and work. I've seen numerous situations where a nice-guy manager let problems slip until they went beyond a simple cure and became unresolvable. Don't let things get to that point.

Nothing but the Truth

Truth is the safest lie.
— Jewish proverb

*The average human being can deal with just about
everything except uncertainty.*
— Robert B. Horton, British Petroleum, America

Good managers never lie. But they've learned to manage
without having to tell the whole unvarnished truth. Instead,
they tell partial truths.

This tactic sounds appalling, but it's not. One of a manager's responsibilities is to shelter his or her people from
organizational mismanagement. And this includes withhold-
ing information from above that could drive the troops to dis-
traction.

Upper management is human, and as such is prone to
the very human condition of making mistakes. Unfortunately, many people unconsciously invest their management
with a bit of the Royal Purple — they want their leaders to
be infallible, to be kinglike. But, hey, company managers
aren't infallible, nobody is. Thus, part of a line manager's job
consists of shielding his or her charges from that reality. If,
for example, you knew of the indecision, the problems at the

upper management levels (most eventually get solved or go away), the uncertainty could paralyze you.

Thus, managers act as a filter, scrubbing out the worst before passing information on to their troops. The question is when does a manager stop, when do their subordinates need the unvarnished truth. The answer: managers need to listen to their own bodies: when such filtering causes unbearable tension, that's the body's way of saying that it's time to tell the straight dope.

The body never lies.
— Martha Graham, dancer

Obstacle Course

No gains without pains.
— Benjamin Franklin (1706 – 1790)
Poor Richard's Almanac (1745)

To most technical people it's a basic act of faith that our organizations use logic and rational analysis to select new ideas and products. Wrong. Most companies employ a hidden mechanism, an internal obstacle course to winnow out and eliminate unsuitable proposals.

If you have a potential winner of an idea for a technology, a new methodology or a product, be prepared to fight for it or just forget the whole deal. Such ideas are not accepted or rejected solely on the basis of rational examination. Instead they, or rather you and your allies, will be forced to run an internal obstacle course, a gauntlet designed to eliminate the faint of heart and those who do not truly believe in their proposals. Only if you truly believe, if you have proven to be strong of purpose and wind in pursuit of your idea, then and only then, will your proposal get a fair hearing.

Ideas are cheap, while company resources are dear. Consequently, most companies have a built-in obstacle course to winnow out ideas and projects. I doubt that these obstacle courses are planned or the result of management cynicism, but rather that they have naturally evolved to fulfill a need. Interestingly, many top managers may not be aware of the existence of such a winnowing mechanism.

Pat Yourself on the Back

Life is an adventure, not a career.
— Graffiti on bathroom wall in Silcon Valley company

Know your own value and be sure to pat yourself on the back now and then for work well done. For as you move up in seniority, you'll get less and less kudos from management.

Junior engineers and programmers get a lot of encouragement. Good work is noted and praised. We all know enough to encourage juniors. Interestingly, we generally do less and less of that for technical professionals as they become more senior. Maybe management figures they know the score and don't need to be stimulated and motivated.

Unfortunately, much of what we know about ourselves is reflected from the folks around us, from their reactions to us and our ideas. These inputs act as a rather distorted mirror for us. And many of us depend on this flawed mirror for our good opinions of ourselves. Thus, it's surprisingly easy for good technical people to actually get depressed or even start to doubt their abilities when they are pelted with criticism, no matter how low-level or ill-considered it is.

So for you managers: try to let your senior people know when they do good work — a project or yearly performance review is not enough. And for you senior professionals, remember the value of what you do is not revealed only in the reactions of others to it. You, after all, know just what you've accomplished and how good it is. Stay on track.

People as Capital

Great people don't guarantee corporate success— but no company can succeed without great people.
— T.J. Rodgers, founder, Cypress Semiconductor

The employer generally gets the employee he deserves.
— Sir Walter Gilbey (1831 – 1914)

People are today's real capital, not machinery, buildings, or equipment. For without the right people, all those means of production aren't worth spit. (Actually 80 – 85% of all business spending is for salaries, although this percentage may be lower for technical firms.)

There is a good news side and a bad news side to this new reality. The good news is that your value as a technical contributor is increasingly understood and valued by your employers. The bad news is that you'd better keep improving your base capital — your knowledge and skills —- or else.

It has always amazed me how management tends to give the benefit of the doubt to fellow managers, no matter how difficult these managers are, or how much damage they do in the lower ranks. As the value of human capital rises, that policy will have to change or companies that don't husband and nurture their most critical resource will find themselves at a competitive disadvantage.

It only makes sense that managers be held accountable for their management of critical resources, especially people. Most companies wouldn't hesitate to punish a manager who destroys capital equipment or carelessly loses money. Yet those same companies hesitate to discipline managers who consistently lose or sour their technical subordinates.

If you're starting out and your company seems to treat its new employees as if they are in Marine Corps boot camp, if the hiring managers seem to devalue working technical professionals, then put some distance between them and you. That is not the technical working world.

Pigeon Syndrome

When a man knows he is to be hanged in a fortnight, it concentrates his mind wonderfully.
— Samuel Johnson (1709 – 1784)

There is nothing so useless as doing efficiently that which should not be done at all.
— Peter Drucker, Management guru

In a behavioral experiment, pigeons were fed by a randomized feeding machine. Over time, each pigeon evolved its own special behavior, a behavior that each erroneously associated with being fed. This behavior is called Superstitious Learning — falsely believing that specific behaviors lead to rewards.

Many companies and organizations show similar — Pigeon Syndrome — behavior. In the last two decades of the high-tech boom, successful companies have associated specific corporate behavior with success. And, like the pigeons, much of that behavior wasn't applicable. Worse, some probably actually worked against success.

These Pigeon Syndrome beliefs are hard to shuffle off. Many corporate belief systems have solidified over time and are extremely resistant to change. "We've always done it this way and it's worked in the past. Why change?" is the typical response. But that reaction is changing as many organizations recast their structures and strategies on their way down to a competitive fighting weight.

Now is the time to root out these Pigeon Syndrome beliefs. Managers are far more willing today to listen to more effective ways of doing things. Companies known for rigidity and constraints are abandoning them to achieve market results. So, if you find yourself constrained by inefficient rules or ineffective methodologies, you have a fighting chance for effective change. Go for it.

I ran across this in "People in the Dark" about Xerox (see references). This story fits in with so much that I've seen, that I just had to repeat it. <<rw>>

Problems

Problems: that's what they pay you the big bucks to solve. A company with problems is a company with opportunity. If things were easy to do, why hire you?

1. Symptoms. Before charging into a problem, make sure that it can be solved directly, and that it's not a symptom that masks a higher level, more complex problem.

2. Simplify. Before you attempt to solve a problem, make sure that you have reduced it to its core constituents. State the problem as simply as possible. Check for unwarranted assumptions. Simplify and then simplify some more. Only when you can state a problem succinctly and clearly can you understand it well enough to create an optimum solution.

3. Conservation. Learn to conserve your energies and attack only a solvable subset of pending problems.

4. Known Problems. Most people know their own problems — personal and professional; they simply don't want to face them, which is perfectly rational. However, given that attitude, there's no reason to be upset over them.

5. Worry. Take the set of all problems that can affect you. First delete those problems that you cannot change, such as the economy and industry prospects, etc. Next delete those problems that you can affect, but choose, for one reason or another, not to deal with. You're now left with the few problems that you can have an effect on and are willing to tackle. These are the only problems worth worrying about.

Product Complaints

Good, better, best
Never let it rest
Until the good becomes the better
And the better becomes the best
— Elementary school rhyme

It cost a lot to build bad products.
— Norman R. Augustine, Martin Marietta
(Augustine's Law #XII)

Complaints are one of the most direct forms of quality assurance you can get. Listen to them. Customers differentiate their suppliers on how well they address and solve customer problems.

Here's a different way to view product complaints. Look at it this way: Somebody is going to a lot of trouble to help you fix your products or create new ones. They don't have to, you know — they can just go elsewhere.

Take complaining about a bad meal in a restaurant. Many view the situation as a test of manhood or womanhood: "Are you going to be a wimp or are you going to stand-up and be a man or a woman and let them know your displeasure?"

A few years ago, after a memorably bad meal, a Unix system programmer gave me a new outlook on complaints. Basically, complaints aren't a test of will at all, but rather a test of whether or not you care about the restaurant and its people. Guess what happens if you and others don't complain? The restaurant will continue to serve bad food and will eventually fail. Not only that, but they won't have a clue as to why customers don't come back and new business falls off.

Sound familiar? Without feedback you can't correct your problems. So value your customers who complain. They could have gone elsewhere and let you sink. Instead they made the effort to help you.

Product Track Record

In high school physics they taught us how to measure work. The formula was force times distance. Take an object that weights 20 pounds and move (raise) it six feet, and you've done 120 foot-pounds work.

... But if you stood and pushed against a wall all day and didn't budge it you hadn't performed any work. Because you'd moved the wall a distance of 0, so it didn't matter how much the wall weighed, the product was zero.
— Matthew Scudder, Lawrence Block
   (Out on the Cutting Edge)

Expending energy and doing work are not the same. If you push a big boulder all day but don't move it, you've expended energy, but not done any work. In physics, as in life, the definition of work includes movement (work = force x movement).
— the author

Doing work means more than just going through the motions or even developing good technology. It means creating successful products. Aim at building a technical track record for your career that includes successful products. Technical experience built up developing products is good; technical experience in creating successful products is more than good — it's great. And managers are increasingly aware of that; they want to see product successes as well as technical skills in their developers.

A dark side to technical work is the fact that many development projects and products never see the light of day.

I know engineers and software developers with over 20 years of development work who never saw their creations make it out of development— the products and systems were killed in development or stillborn.

For technical people there is a constant temptation to work on the high-tech flyers, to tackle products that embody risky but great technology — and corresponding high failure rates. Indulge yourself, but remember to have some successes as well. Developers who have worked on a long string of product failures can end up as a modern, technical equivalent of a Jonah or a Hard-Luck Harriet. The thought is: put them on a project and it's a sure failure.

It's hard to keep the technical faith through a series of technical or product defeats. And faith in technology and ourselves is what drives us. Remember, while the world values a competitor, it loves a winner.

Project Management Software

You can never plan the future by the past.
— Edmund Burke (1729 – 1797)

Project management software is a good example of idealized software that doesn't match the real world.

Today's project management software packages are descendants of PERT* and CPM.** These are methodologies developed for controlling large projects such as developing the Polaris missile or building major plants. Currently, there are a number of software packages out there for planning and scheduling projects. In these packages each project task is scheduled and integrated with others; the software calculates worst case paths and automatically updates scheduling for changed task times for the entire project.

Unfortunately, such project management software represents an idealized view of work (not the real world). Managers in real life play with at least three sets of numbers: the schedules submitted by their troops, the schedules that managers believe are realistic, and the numbers that they submit to their management. Unfortunately, no project management software package allows users to easily juggle these three or more sets of numbers in a single project.

Additionally, project management software doesn't deal with normal technical project conditions. For example, there's no mechanism for automatically rescheduling a task. And there's no pass/fail test for automatic rescheduling, nor a mechanism for automatically adding additional tests as needed.

It would also be nice to have multiple projections or representations based on levels of optimism. For example, that would include projections which are: optimistic, pessimistic, plain vanilla (mainline), and Goldilocks (just right — the optimum solution).

* PERT— Program Evaluation and Review Technique
**CPM — Critical Path Method

Promote Your Boss

Working for someone is a lot easier than working against them.
— anonymous

If you want to get promoted, help get your boss promoted. Books and movies love controversy and confrontation. A favorite tale is that of the underling who tunnels under his boss, displaces him and gets promoted into his boss's slot. In the real world, trying to knife your boss in the back is a fast way to earn a one way trip to the unemployment line.

Good management won't promote a manager unless that manager has trained a replacement. If you want your boss's job, make him or her look good, help the department succeed, and get your boss to train you as his or her replacement.

Promotions Are Not Just Rewards

We pay for performance and promote for ability.
— A major insurance company's compensation strategy

Promotions are generally not rewards for past performance. Instead, management uses promotions to advance those who display the potential to tackle the next level of bigger, tougher problems.

When most of us do a good job, we expect a reward, a promotion, or a raise. Yet good management is forward looking: they reward us not just for what we did, but for what we are going to do next. The situation is a variation on: "What will you do for me next?" Not: "What have you done for me lately?"

So if you want to get ahead, do spectacular work. But also aim for the next set of higher level design or management problems. Let your management feel that promoting you will be a good investment. Don't live on past glories; instead live for future ones.

Doing a good, competent job is not enough for a promotion or out-of-ordinary reward. If you want a promotion, then you have to show the capability for doing more and handling tougher tasks.

Respect

I get no respect.
— Rodney Dangerfield

Don't work for someone who doesn't respect your skills and capabilities. If you do, sooner or later you'll suffer for it.

Don't wait it out. Either change your manager's opinion or get out — get another job or find another boss in your organization or company. Unfortunately, such situations usually don't get better. Working for someone who does not professionally respect you is like sitting on a time bomb; sooner or later it will go off.

Remember, you only have a certain amount of time in this life to accomplish things, to build your skills and to grow. Try to be as productive with this time as you can: you'll get more done, feel much better about yourself, and have a better life. You simply can't do this while working for someone who doesn't respect you; they'll undercut you and restrict the scope of your playing field. You're going to lose.

If you have a track record of respect-type problems, of working for people who don't give you professional respect, then you may be the problem — after all you are the common denominator. There are a couple of choices. Either

1. You're lousy at picking jobs, or
2. There is something about how you act that causes you to be mistreated or suffer disrespect.

In the first case, learn to be more discriminating about potential jobs and managers. Pay attention to how managers act; look for verbal clues; and ask others working at your level how things really are.

In the second case, you probably need to be more self-assertive. If you have and radiate self-respect, you're on your way to getting professional respect (don't forget you still have to perform). Those who insist on professional respect generally get it (or the door).

Results Versus Personality

Virtue is its own punishment.
— Anonymous

Technical people differ from most businessmen. We made a pact early on — an agreement to be judged not on our personalities and relationships, but rather on the results of what we do.

Because of this unspoken compact on results versus personality, managers should be extremely careful in promoting or rewarding people. If people are perceived as being rewarded on a personal basis, rather than for technical accomplishment, it can result in disastrous consequences; the best technical people will feel double-crossed and leave.

Technical people spend years building up their skills; competence is a matter of understanding technology and being able to deploy it to implement solutions. Winning on personality, not on a technical basis, has been a poisoned career apple for many a well-rounded engineer or programmer. It can become a habit that inexorably leads to professional suicide. It is, after all, far easier to paint word pictures than to actually do something.

So beware. Don't get on the personality merry-go-around. Personality can help, but remember first and foremost to deliver results.

Reviewing Subordinates —
Advice for a First-Time Manager

The trouble with personnel experts is that they use gimmicks borrowed from manufacturing: recruiting, selecting, indoctrination and training, job rotation, and appraisal programs. And this manufacturing of men is about as effective as Dr. Frankenstein.
— Robert Townsend, AVIS

If you're an experienced manager, skip this, it will probably infuriate you. On the other hand, if you're a first-time manager this may help you keep your priorities straight, especially when it comes to salary reviews for subordinates.

As a first-time manager, you are constrained by a number of company guidelines, and probably feel that you're being judged by higher management on just how well you adhere to these strictures. But, there is more at stake than satisfying rules. If your people don't deliver good work, it won't matter what rules you follow or don't follow. Ultimately you are judged by your subordinates' performance.

Therefore, it pays to be a bit generous to your people; it pays to violate a few rules on the side of the angels, not the accountants. If you can motivate your people, if they believe that you're on their side, if they deliver, then they will make your reputation. On the other hand, if you can't motivate them and get them to perform, they can sink you.

Rules

Rules are to keep the idiots from screwing up.
— JPL engineer

Rules are yardsticks, not crutches.
— Gene Olson

Every dogma has its day.
— Abraham Rotstein

Rules can be the death of an organization.

The human mind did not evolve to sit in an armchair and philosophize; instead, it does best at recognizing the tiger lurking in the brush and shouting "jump." To quickly recognize patterns, we humans generalize, we automate actions to simplify and speed response.

Similarly, rules have evolved in the body corporate to provide quick, standard answers to common operational problems. Unfortunately, these rules can outlive the problems they were put in place to solve. When that happens you have the means — the rules — dictating ends. This is an intolerable situation: it can lead to gross inefficiencies.

The difficulty in getting rid of archaic rules is that they get built into the warp and woof of the corporate fabric and become gospel. And once a rule is accepted as gospel, it's almost impossible to have a rational discussion that questions its usefulness. Defenders of the status quo will endlessly quote the rule chapter and verse to defend its existence.

Here's a swift, modest proposal to root out dated, obstructionist rules. Instead of arguing the pros and cons about a rule, use a customer or a user as a guidepost. Simply ask them what they think of the rule under discussion. In many cases, you'll get a reaction on the order of: "You've got to be kidding, you're doing what?" If the rule doesn't make sense to them, take a good look at it; the odds are you have stumbled on a relic that deserves the deep six. Kill it.

Running Lean

One person with only half a job can wander around and do real damage in his or her spare time.
— Robert Townsend, AVIS

El que no tiene que hacer, piensa en los males que hara. (The person who has nothing to do, thinks about the mischief he will do.)
— Mexican proverb

Small is not only beautiful; it is powerful.
— T.J. Rodgers, founder, Cypress Semiconductor

Running lean makes for an effective organization. With too few people, everybody cooperates and hustles to get the job done. Folks don't have time for politics, personal biases, or back-stabbing. When running hard we cut corners and do what we must to move product.

On the other hand, organizations that have more people than work tend to have world-class office politics. The people have the time to start envying others: she has more office space than I do; he has a better desk, they have less work than us, etc. Makework multiplies, clogging the organizational ways. And boredom sets in.

Idle hands can destroy organizational consensus faster than layoffs. Someone should do a study on the number of memos generated as a function of manpower loading. I'd bet dollars to doughnuts that the less work folks have to do, the more makework memos they churn out.

If your group or company has too many people, look out. If it gets bad, look in the organization for another place that's running lean, a place where you can get things done without hassle.

Sacrifice

High places have their precipices.
— Barber
(The book of 1,000 Proverbs, 1876)

It is better to be the reorganizer than the reorganizee.
— Norman R. Augustine
(Augustine's Laws)

Managers, especially high-level ones, serve another, special function. They are the ultimate sacrifice, a pawn to get company-wide attention.

Suppose a company or division or group has to change its ways, to get efficient or die. What's the best tactic to get everyone's attention? One way is to fire 5% or more of the staff. Another is to simply fire the head man or woman: that action gets everyone's attention and only involves dismissing one person.

Managers, you see, also serve a symbolic role, as a quasi head-of-state. And sometimes, for the good of that state, their heads will roll as a symbolic sacrifice, a scapegoat.

Interestingly, many managers don't recognize this part of their job. It's a surprise for them to discover that they're expendable precisely because they are so prominent. And that they hired in not only to lead, but also to serve as a convenient scapegoat.

Salary/Performance Reviews

Understanding should precede judgment.
— Louis Dembitz Brandeis, U.S. Supreme Court Justice
(Burns Bakery vs. Bryan, 1923)

It all comes back to rewarding outstanding performers. Great people expect to be rewarded. You can't reward great people unless you identify them fairly and accurately.
— T. J. Rodgers, Cypress Semiconductor
(No Excuses Management)

Deal with salary/performance reviews early and make sure that there are no surprises for either you or your managers.

Nobody likes Salary and Performance Reviews. From the manager's point of view it's a no-win situation; somebody is sure to be unhappy. As for technical performers, reviews never seem to compensate them for the energy and dedication they pour into their work. Salaries will continue to be inequitable between people, due to differing starting points for people and the constraints of the salary system.

Yearly salary and performance reviews take on the form of highly ritualized morality plays, plays in which no one wins. Reviews are where most folks find out just how their perceptions and expectations differ from those of their reviewing manager. Unfortunately, by then it's too late to change anything. At that late date, the paperwork is already filled out and it's exceedingly difficult for the manager to redo the paperwork or admit that he or she made a mistake.

Head this situation off. Make sure that there are no real surprises at your yearly review: talk to your managers early. Let them know what you think and want; and try to find out their views and what they intend to do. (And remember this if you become a manager.)

Some hints on handling your review:

1. Make sure that there are no surprises.

2. Talk to the reviewing manager at least 8 weeks before your review. Make sure that he or she knows what you think and expect. And try to ensure that there aren't any major differences between your view and the manager's; if there are at least you both know of it.

3. Track your time; see where it goes.

4. Track the problems that you've solved, both for yourself and others.

5. Track your schedules and projects.

6. Let your management track your progress and what problems are developing.

7. Update any schedules and disseminate them whenever your project schedules are affected by external events, such as requirements and implementation limits or changes.

8. Talk periodically to your managers, especially the one who gives the review. If there are any problem areas you need to work on, make sure you have the time to correct them before the review.

9. Make sure others, including the managers, know of any technical triumphs that you pull off. Don't assume that everybody knows.

10. Bring most of this supporting documentation with you for the review.

11. Always try to correct any misunderstandings or erroneous perceptions of you held by your manager. A

small misconception can't hurt you, but a number of them, gathered over time, can.

Sales

Everyone lives by selling something.
— Robert Louis Stevenson (1850 – 1894)

Engineers, programmers, and technical managers tend to resent marketing and sales people — especially their higher compensation rates. That's dumb. Without sales your company is history. Products or services don't move unless there are salesmen and saleswomen who get out there and push their use.

Technical people live relatively stable organizational lives. In contrast, sales folk live on a knife edge. If your project fails, your management doesn't automatically can you. However, sales people either make their nut — their sales quotas — or they're history. How would you like to wake up every morning with that "produce or die" sword hanging over your head?

Years ago I remember going into stores where a row of salesmen would sit against the wall like sunbathing alligators, lazily awaiting their next meal. Wrong image. Salesmen/

women are the keys to moving products through distribution networks and into customer hands.

And by the way, you had better learn to sell as well, to sell yourself and your ideas to your management. No product succeeds without some form of salesmanship, and that includes you.

The Sandwich Technique

I praise loudly; I blame softly.
— Catherine II (the Great), Russia
(1729 – 1796)

Nobody likes criticism, let alone listens carefully to it. One effective way to deliver criticism is to package it between praise — referred to as the S--t Sandwich Technique.

Managers don't criticize performance or designs because they get a high from picking on people. They do it to improve performance and results. Unfortunately, most people automatically drop into a defensive crouch when criticized, rather than listening and then acting on it.

Criticizing people or their designs is a tricky bit of business. The sandwich technique is one way to defuse or delay the defensive reaction. Lead with honest praise for something, then follow with the criticism and end with more honest praise. The mix disarms folks, bypasses their defenses, and opens them up to listen to what you are saying. It also makes them feel that you're paying attention, not just looking for negatives.

An example would be: "The XXXX idea is a good one, because... But here are some problems we have to fix to make it work effectively... However, the idea base is solid and I think it will succeed..."

This technique works, but you can't trivialize it into a "One Minute Manager" tactic of a one-two-three, wham-bam-wham delivery and then you're gone. You have to take the time to review both the positives and the negatives. And you have to be honest: praise the good stuff and criticize those aspects that need it.

Be careful to reserve this tactic for important issues. If you apply this approach to everything, people will learn to expect it and simply wait to hear the bad news.

SOBs —
Working for an SOB

I don't get ulcers, I give them.
— Harry Cohn, film producer

Working for an SOB can actually be better than working for a nice guy or lady. At least SOBs let you know exactly where you stand — they don't beat around the bush. Whereas, nice guys tend to avoid confrontations. Nice guys sometimes put problems off too long, to where minor problems turn septic and become unsolvable. SOBs, on the other hand, don't hesitate, they're on it right away. Before you know it there they are, in your face and saying what they think.

In dealing with SOBs you have to set your feet firmly and stand your ground. If you can't tell them where to get off, you're probably better off not working for an SOB. But, if you know your stuff and can stand up to them, you can get the leeway needed to be effective.

Most SOBs are not used to being called to account on their behavior and many, believe it or not, actually handle it rather well. Just be sure when you take a stand that you're right and know what you are talking about. Bluffing doesn't work with SOBs.

There are, unfortunately, some totally irrational SOBs, who are impossible to work with. With them nothing matters; you'll be in trouble whatever you do. If you get stuck with one of these, you might as well find another job.

Solutions Not Problems

Opportunities are usually disguised as hard work, so most people don't recognize them.
— Ann Landers

A violent plan executed today is far more effective than an excellent plan executed in a week.
— General George Patton, U.S. Army

If you want to get promoted, bring solutions, not just problems, to your boss's attention.

A number of folks just can't wait to run to their boss with the latest problem. By doing that, they feel noticed and important. Unfortunately, most managers are already awash in problems; they don't need more problems —they need solutions.

So if your managers will be grateful for your identifying a potential problem before it bites them you-know-where, they'll be overjoyed if you actually bring along a proposed solution as well.

Additionally, you'll have a solid advantage in bundling your solution with an identified problem. You're presenting the first solution, and if it's an effective one, you may not have to run the usual gauntlet against competing solutions — you can just go ahead and implement it.

Standards Committees

*There comes a time in the affairs of man, when he must
take the bull by the tail and face the situation.*
— W. C. Fields

*Committee — a group of men who keep minutes and
waste hours.*
— Milton Berle

Generally, standards committees are a sinkhole that sucks
in people, resources, and time, while producing meager
results. Most technology becomes an industry standard for
one of two reasons: one, like the PC, because it's the de facto
standard and dominates its market; or two, by default, as
the surviving, obsolescent standard.

I don't mean to be cruel; we need standards and we actually have some committees doing outstanding work on key
standards such as those for SPARC, SPEC, and CALS. But
most standards take years to emerge, and are usually too
late to serve as effective industry guidelines for designers.
Moreover, standards groups seem to attract a standards type
of person — one who is comfortable with a snail's pace to
reach standardization.

I once conceived and helped set up a standards committee,
which was highly successful — SPEC (Systems Performance

Evaluation Cooperative).** SPEC, a nonprofit organization, defined a set of application benchmarks to be used for measuring high-performance desktop/server computer systems.

SPEC set standards history: its first benchmarks (10 applications running on Unix operating systems) were out and running in one year from the first meeting. (Probably because the first meeting was held in an Irish bar — Stanley's in Campbell, CA.) SPEC benchmarks are now the de facto standard for measuring microcomputer performance and are used by all the microprocessor vendors.

Actually, SPEC succeeded because it had very good committee members, a minimalist philosophy (enforcement by users), and started out with only a few member companies (Sun, HP, Apollo, and Mips). The small initial organization set an effective tone for later members.

SPEC Lessons

1. The fewer initial members the better.
2. Committee member companies must commit resources for internal work to support the standards development.
3. Self-enforcement works best (don't be a policeman, let the user community do the policing).
4. Link in the press (it keeps things honest and provides for immediate coverage).
5. Minimize paperwork. Keep it simple.

**Systems Performance Evaluation Cooperative (SPEC)
c/o National Computer Graphics Association
Suite 200
2722 Merrilee Drive
Fairfax, VA 22031
(703)698-9600 extension 318
(703)560-2752 FAX

Stop & Think

Every child knows that prevention is not only better than cure, but also cheaper.
— Anonymous

You can solve more debugging** problems by stopping and thinking than you can by continually rerunning or resimulating the design and trying different fixes.

Debugging (finding errors) can be an activity trap. The temptation is to try again and again: to instrument it differently; to run the code or hardware again to catch the error; to put in a temporary fix and try again. This approach is very seductive these days because it's so easy to simply run another iteration.

Unfortunately, you can end up fixing minor problems, or worse, spreading errors by implementing temporary fixes that themselves cause other problems farther down in the execution. And you can get tired — it's a lot like long distance running or swimming, you just keep putting down another step or stroke. Continual retrying becomes a rote activity: you've turned off your brain and are now on automatic pilot.

While a digital hardware and software designer, I always found that I could solve problems much faster by going off and thinking about them, rather than keep dynamically running the code or hardware and trying to grab error symptoms and tracing them back to the error sources. Maybe it was just my lack of brainpower, but a lot of folks seem to have had the same experience.

It's far better to find errors early. Just sitting down and rechecking your design can have tremendous benefits, as do design reviews. In software too many programmers concentrate on monitoring the executing code threads; they should pay more attention to the underlying data — many times it clearly highlights error conditions.

** Debugging — the process of finding and eliminating errors, known as bugs. Tradition has it that an early computer failed because it had a real bug in it that shorted a line. Ergo — a bug is a hardware or software error. And "debugging" means getting rid of bugs.

Take Care
of Support People

You can't do it alone.
— Anonymous manager

Light is the task when many share the toil.
— Homer (700 B.C.)
(The Illiad)

Take care of your support people and they'll take care of you. Don't be domineering or arrogant to secretaries, technicians, or administrative people. They can help you or they can bury you.

While in college I ran a printing press for a county department. To the nice and friendly customers with printing jobs, I was as helpful as could be; I made sure that their jobs ran right. But for the "do-exactly-as-I-say-and-don't-talk-back" crew, I did exactly what they asked; I had no choice. Needless to say, they got back some pretty crummy work. They'd given me no leeway, no way to help them.

Difficult people don't understand: you can't do it all yourself. And worse yet, sooner or later you'll screw up. Push people around and they'll have no commitment to save you from your own mistakes — they'll let you drown in them. To succeed you need to let people help you, especially those in supporting roles.

Take the High Road

Always do right. This will gratify some people and astonish the rest.
— Mark Twain (1835 – 1910)

The trouble with the rat race is that even if you win you're still a rat.
— Lily Tomlin

Take the high road when you have a problem with people or policy, or have things you want changed or done. If you want to win, the trick is to present your proposal in terms of a higher good, of benefits to the organization itself, and not just to yourself.

Most managers are sick to death of people-vs.-people problems — the "I said this, she said that" kind of thing. And they hear "I want, I wish, I need, I feel…" all the time. Even the best of managers eventually become jaded with such verbal pyrotechnics and automatically turn off when faced with such truck. They listen, but they don't hear.

So if you want a full hearing, if you want to get your way, if you want your appeal to succeed, then you must address a higher level. Instead of personalizing your needs, put them in managerial terms, clothed in the rational needs of management. Suppose, for example, you need a more powerful

193

computer. Instead of demanding it as a personal reward, make the case (assuming that it's true) for upgrading the department or analysts at your level, and show the organizational benefits of doing so.

Or suppose you have a confrontation with someone on a policy or implementation issue. When you bring your case to management, don't make it personal. Instead, take a higher level track, appeal to the goals and perceptions of management. Try to solve the management problem directly, and yours secondarily if at all.

Talk

Syllables govern the world.
   — John Selden
    (Table Talk)

Woe betide you if you're an engineer or programmer and your management discovers that you can talk to people and make sense. And if you can also manage or sell, then you are on your way out of technical design.

Technical disciplines are woefully short of those that can deal with people effectively. Thus, anyone who combines technical capability with people skills will find themselves quickly dragooned up into management, product management, sales, or marketing.

Once discovered to have these skills you'll have a tough battle to try and stay technical. Be prepared for a fight if you want to stay in design.

Thinking

Few people think more than two or three times a year.
I have made an international reputation for myself by
thinking once or twice a week.
  — George Bernard Shaw (1856 – 1950)

Thinking is the hardest work there is, which is the
probable reason why so few engage in it.
  — Henry Ford, Ford founder

Thinking....That's what you're paid for. Surprisingly, most of us are uncomfortable with actually buckling down and deeply examining something. Few of us do it. We generally find makework to do instead of thinking.

Shaw's right. In my case, I can occasionally do well thinking about problems and solutions. Do I do it often? No way. I have to be trapped by schedules or circumstances before I'll buckle down and really think about something. Similarly, I've found writing to be easy, but the required thinking to be hard. So I delay writing to the last minute and face the thinking part only when I must. I suspect that this aversion to productive thinking is a common human trait. To most people, action seems to be far more preferable than thinking.

I find that I actually do my best thinking in boring meetings and presentations. There one is trapped: There's nothing else to do. You can't read, or draw cartoons, or fiddle with toys, or water the plants, or even sleep. Boredom drives you to the last resort — thinking.

Maybe you can develop the skills to sit down and think. Take the distractions away and examine the problem carefully. Don't jump too fast to a solution, that's the easy way out. Turn the problem over; go at it in different directions; ask new and different questions. There are just not that many people willing to sit down and think. If you can develop the skill, you can really contribute. Try it.

Trading Up

The future of work consists of learning a living.
— Marshall McLuhan

Avoid the grunt work trap. Everybody has to do grunt work now and then, but if you have to do a lot of it make sure you trade up — swap grunt work for a chance at future skills.

It's unfair to rely on your manager to watch over your career. They have to juggle what's good for you against what's good for the organization. Managers always have low-level but important work that needs doing. If you have to do it, fine, but make sure it's in trade for better, skill-building work downstream. Talk to your manager and say: "I'll do this for you now, but next quarter you let me work on...."

Many engineers and programmers get caught in the "I'll do what's necessary" syndrome, counting on their managers to even things out. They end up doing important grunt work on a permanent basis. Unfortunately, their salaries keep increasing, but not their skills. At some point they no longer earn their keep and become vulnerable, especially if they have a new manager.

Do what needs doing. But the best course for both yourself and your company is to continue to build up your skills and technical worth.

Travel Tips

I like terra firma — the more firma, the less terra.
— George S. Kaufman (1889 – 1961)

Travel is now a constant part of the technical life. Here are some hints to ease the burden:

1. On the West coast stay in small motels instead of big hotels. You can park in front of your room, check in and check out in 1 minute. They are also dirt cheap. And you can go in and out to use the phone and bathroom.

2. Always keep your ticket in the same place, in your Day-Timer or schedule book or a trip folder.

3. Keep a manila folder for each trip, with your ticket in it as well as your itinerary and all expense receipts (staple an envelope into the folder to hold them). This approach makes it easier to do expense reports.

4. If you're visiting a number of companies, ask them to send you a confirming fax with directions and a list of who you're meeting. Staple that into the trip folder.

5. If the hotel or motel doesn't have in-room computer connections, you can always hook up in the business office or at the front desk via their fax line (just do it late at night).

6. Checking-in airline luggage is OK, except if you have any connecting flights. Then the inevitable will happen and your luggage will be lost.

7. Calling credit cards are great — you don't have to keep dialing in your PIN number.

8. Take a small magnifying glass for reading street maps in dim light.

9. Beware of connecting flights with less than 40 minutes between flights, especially if the hub is a place like Boston.

10. Have your travel agency get seat assignments with your plane tickets. This saves time, you can go right to the gate and check in (you can check in any non-carryon luggage at the curb).

11. Generally if you order a special meal on a flight, such as low-fat, you will be served early.

12. If you go to seminars or conferences and end up with a lot of supplemental material, you can FedEx the stuff to your office and not have to lug it onto the plane. Most conferences and hotels have shipping facilities.

13. You don't have to eat in the hotel you're staying at or have to take your chances with the surrounding restaurants. Ask the doorman, concierge, or clerk for a good place to eat. They know and will be glad to tell you where the best is to be found.

14. If you're attending a conference or seminar that requires a lot of walking, you don't have to suffer painful feet. Get yourself black sport shoes like Reeboks.

There are even versions for men and for women that can pass as dress shoes.

15. If you have a lot of luggage or cases and have a rental car to return, you don't have to lug that stuff around. Drive to the terminal first and check in your bags at the curb before returning the car.

Treated as You Expect to Be

No one can make you feel inferior without your consent.
— Eleanor Roosevelt

You're treated with the respect that you expect, no more or no less. Demand respect and you will get it or the door. But you will not be treated with disrespect.

In my software days, I went to work for a small system software company. On my first day, the VP met me and took me to see the president. On the way, he warned me that the president had one quirk: he really hated anyone to use the word "irregardless." "So," he said, "whatever you do, don't use the word 'irregardless.'" Of course, as soon as I met the president (a great guy) I intentionally used the word "irregardless" two or three times straight out. The result: they both laughed, and that was that. I figured that if there was going to be any craziness, it might as well be right then and there. Why wait? And of course, as these things generally go, everything worked out fine.

If, like Rodney Dangerfield, "you don't get no respect," you'll be unable to do really good work. Thus, sooner or later, your neck will be on the chopping block. So go ahead and push for the respect and the independence needed to do the job; that's the best job insurance you can get.

Turning an Oil Tanker

It takes 5 years to develop a new car in this country.
Heck, we won World War II in 4 years.
— H. Ross Perot

Turning an oil tanker is child's play compared to turning around a corporation or an organization. Most technical people become terribly frustrated at the slow pace of internal organizational change. Many simply give up and job hop, unwilling to stagnate while waiting for change to finally take hold.

Organizations just don't change easily. A good analogy to the delays associated with major change is that of a supertanker. Picture this: there you are on the bridge and give the order, the helmsman spins the wheel to port, and 3 miles later you start the turn, finishing it in another 10 miles or so. It takes time. So do organizational changes.

So be patient on long-range change. Instead of giving up hope, concentrate on faster, short-range changes. Work to have your management accept continuous, incremental improvement. You'd be surprised at the cumulative effect of small, incremental changes (so will your managers).

Two Kinds of People

Amateurs hope. Professionals work.
  — Garson Kanin

*I am only an average man, but, by George, I work harder
at it than the average man!*
  — Theodore Roosevelt

There are two kinds of technical people: them that do whatever has to be done to get a job finished, and them that don't. It's that simple. Some people get things done, others don't. Some consistently perform, some always have excuses.

In the technical world, it's results that count; and in most cases those results are highly visible. So to succeed in this technical arena, be sure that you finish what you start, and that you do what you say you'll do.

Performers are not necessarily the brightest or the most impressive in meetings. But they are known by what they do. On the other hand, many technical people show great initial promise, but simply fade in the project home stretch.

So if you want to do well, be sure that you end up a performer, not a laggard. Set your priorities to do what has to be done. Don't fall into the easy trap of not doing extra work for the company. You're not — you're doing it for yourself. The most valuable thing you possess is your technical skills, as well as the will and drive to wield them. Never, never let anyone or any company destroy that — it's the only real job security that you'll ever have.

Use Your Boss

Advice to a Manager:

Give up being an administrator who loves to run others and become a manager who carries water for his people so they get on with the job.
— Robert Townsend, AVIS

He or she is the best resource you have. If you get things done, he or she will be glad to help you to be more effective. Use your boss to get you the resources and playing field that you need to meet your schedules.

Many technical people look on their manager as a top-down authoritarian, as someone best dealt with at a distance. That's foolish.

Your manager succeeds if you succeed. If you have a good boss, he or she will know that fact of technical life. So use your manager and his or her clout to get you the resources to get the job done.

The Vanishing Manager

Far and away the best prize that life offers is the
chance to work hard at work worth doing.
— Theodore Roosevelt

Don't count on loading your managers down with support tasks; instead, you may end up doing their work. These days technical professionals are taking on more and more of what used to be management's role. For better or worse, they are on their way toward self-management. Being your own mini-manager is a major consequence of today's move toward flattened organizations.

You can take either the glass half-full or the glass half-empty view of this self-management trend. On the glass half-full side you get the freedom to decide, to act without bureaucratic delay. In essence you will be what engineers and programmers always wanted to be — a professional, with the responsibility to make key decisions and the accountability for results.

With the glass half-empty point of view, you end up taking on many tasks that managers previously did for you. But

— and this is a big but — you won't be compensated for this extra work. To some this means that you're being taken advantage of, and working harder for no additional compensation. Yes, you're freed up to do what must be done, but you're not paid extra for taking on more responsibility. For most technical people this situation is a major win. In my experience, the weight of management inertia and compartmentalization costs people far more in frustration and wasted time than self-management will cost in extra work. But don't trust me: think back to what most frustrated you in technical work; nine times out of ten the cause was inappropriate or delayed management decisions. This shift is a net win.

More self-management doesn't necessarily mean that you'll be left to stew in a solitary world of isolation and your own mistakes. You can band together with team members and collectively address many problems previously handled by a manager. Management is still there to provide access to company resources (i.e., funds), to help out as needed, to review projects, and to reward or punish behavior. You just have more freedom to decide and act.

Walking the Parking Lot

How is it possible to expect that mankind will take advice, when they will not so much as take a warning.
— Jonathan Swift (1667 – 1745)

As a grunt and later a manager, I spent lots of time in what I'd call "Parking Lot" walks — walks around the parking lot solving personnel problems.

As a grunt I was the problem child, either getting chewed out for bad judgment by my manager or complaining to him about some idiotic constraint or problem. As a manager, it became my turn to walk the lots with problem children and to try to solve or at least to defuse tense situations.

Like many designers, I was intense and driven — I wanted to get things done, to move logic and code. And I was frustrated at conditions or people that either hindered or distracted me. I'm sure that many of you are or have been in a similar spot.

However, on looking back, I think I spent too much time walking those lots. It was generally a waste of time: things either got better or they didn't. (We would do an outstanding

215

job or not; the design tides would come in and the tides would go out.)

Nowadays my advice is simple: get in there and do what has to be done. And try to ignore the people and organizational problems — they'll get resolved. And cut your parking lot time.

By the way, things haven't changed: when I visit companies I often see troubled pairs of designers and managers still walking the parking lots as in days of yore. Better them than me. (Come to think of it the last time I saw my boss, we took a walk and...)

When Solving Technical Problems Isn't Enough

Difficulty is a severe instructor.
— Edmund Burke (1729 — 1797)

Engineers and programmers are trained and rewarded for solving technical problems. Starting in school, all that is asked of them is to solve the problems at hand. If they reach the right answer, then all else is forgiven.

Later in industry, again, all they have to do is solve the design problems tossed in front of them. Come in late, don't wash, never go to meetings — things are still O.K. as long as they continue to deliver workable solutions.

But somewhere on the road to middle age and working at a higher level, a funny thing happens — the technical solution is often not enough. More is needed: solutions that encompass technical, people, marketing, and organizational components. And then many a high-cruising techie hits extreme turbulence. They are shocked: the technical solution doesn't work anymore.

This situation can be especially crushing for aging technical professionals; many can't let go of the technical solution. Instead of changing, they act like a windup doll, and keep walking into the solution wall, unable to adapt. Be aware that as you rise up the technical management ladder, the problems you'll solve won't be purely technical. And, in many cases, hard-core technical solutions need not apply. Be prepared to craft wider solutions.

When You Are One...
I'll Make You One

We have no difficulty in finding the leaders: they have people following them.
— William L. Gore, CEO, W. L. Gore & Associates

Good followers do not become good leaders.
— Laurence J. Peter, Author of *The Peter Principle*

Management will make you a manager or leader when you're already functioning as one. The true test for management capability is whether or not you take responsibility when necessary and get things done.

Many engineers and programmers wait patiently for management to anoint them as managers before they'll make the effort to act as a manager. And it will be a long, long time, if ever, before they are ever selected for management.

Act like a manager — take responsibility and produce results. Do that and you'll be made a manager. You can't wait for lightning to strike; you have to take the initiative.

The best managers I know got into management not because of ego, but because they wanted to get things done. They recognized the need to work at a higher level to be technically effective. Also, many felt a strong sense of obligation toward their fellow workers and projects. And they took the appropriate action long before they were promoted into management.

Work at Your Salary Level

You know what happens in the beehive? They kill the drones.
— Congressman William Poage

Growth is the only sign of life.
— John Henry, Cardinal Newman

Work at your salary level or else. If you don't work at your salary level, if you don't deliver work that matches it or is higher, then it's only a matter of time before you're history.

This is a simple, common sense idea; yet it's surprising how many technical people (and managers) ignore it. If you want security, then be cost-effective for your employer. (You'll feel better, too.)

If you're not delivering at your salary level, then change what you're doing. Increase your output or look for more important, valuable work and then do it. Don't count on past performance to excuse working at below your cost.

Make it part of your career planning to check on the value of the services you provide your company. If lower salaried folks are doing the same thing, that's a strong hint to upgrade your work.

There's no reason age should disqualify competent technical professionals. That is, provided that they earn their pay. If you're paid more, be sure that you deliver more. If you don't want to take the management track, one way to increase your value is by working as a technical lead and directing a design team, as well as training newer technical team members. Another is to work with customers.

Writing Memos, Letters

Nobody likes to write, but everybody likes to have written.
— Publishing proverb

If you can't write a good memo or letter, you have a handicap that can limit your effectiveness. Ideas have no currency if others cannot understand and make use of them.

Writing a good memo or letter is not that hard. Thinking out what you want to say: THAT is the hard part of good writing.

Keep memos highly organized, take a pyramidlike approach, with concentrated concepts at the top, working down to massive detail on the bottom. Make sure of your facts and don't assert anything that you cannot, if asked, prove. A memo is no place for opinion. Make your case and do it at a high level.

Keep letters short and to the point. Put your main thoughts on the first line of each paragraph, instead of at the end as your high school and college English teachers preached. Putting them first lets readers skim and still get the gist of what you're saying.

For both memos and letters:

1. Keep it simple (KISS — Keep It Simple Stupid)

2. Don't use cliches.

3. Don't use standard, meaningless forms ("In reply to," "as per your letter," "sincerely yours," etc.).

4. Start out in a direct manner, don't beat around the bush. Say exactly what this is about and then detail it.

5. Summarize up front, not at the end.

6. Use friendly language where appropriate. Contractions such as we're, I'll, etc. are less formal and more readable.

7. For foreign correspondence keep your text simple: don't build complex sentences and minimize the use of contractions.

I actually enjoy writing memos, letters, and E-mail. Why? I just wait until there's something that I really don't want to do, some hard, tough task. I then substitute writing for it — then writing is a piece of cake in comparison. It's like taking a refreshing break. Try it that way — you might find that you like doing memos too.

You're Known by the Problems You Tackle

Del dicho al hecho hay mucho trecho.
(There is much distance between saying and doing.)
— Mexican proverb

Talkers are not good doers.
— William Shakespeare (1564 – 1616)
(Richard III)

Talk is cheap. In the long run it's what you do, not what you said you'd do that counts. You will be judged on the problems that you tackle.

The easiest way to judge someone, whether a manager or a technical performer, is on the size and scope of the problems they tackle AND complete. Many people talk one role, but act out another, lesser role.

One of the most revealing things about anyone is the kind of tasks they instinctively turn to and do. People do what they're comfortable doing. Simple.

Thus, if you want to move up the technical and managerial ladder, the technique is simple. Just take on larger and larger problems, and complete them.

You're Working for Yourself

Make every decision as if you owned the whole company.
— Robert Townsend, AVIS
(Up the Organization)

There are no traffic jams when you go the extra mile.
— Anonymous

You're not working for your boss; you're not working for your department; you're not working for your division or company — you're working for yourself.

1. It's your career that's at stake.

2. Everything you do or don't do will contribute or detract from your professional reputation.

3. Remember, if you're a junior or intermediate engineer, that your career will last longer than the people who you report to. Thus, you have more future time in the business at stake than they do.

4. Whatever you do has two effects: one, on how it helps your company reach its goals; two, on how it adds to your reputation and skill base (marketability).

5. You are more than a janitor: you are a technical professional. As such, you have a commitment toward

professional excellence. A schlock job can have a longer range effect than to just lose money for your company; it can significantly cut future opportunities for you either in your company or industry.

6. You've already made a large investment in your craft through school and work; you want to protect that investment, both for yourself and for your employer.

7. What's good for you professionally should also be good for your employer. There is no dichotomy or conflict. Good professional work is a win-win affair for all concerned. So do what is professionally called for, no less.

Your Worst Enemy

No man is demolished but by himself.
— Thomas Bently

Your own worst enemy is, guess who? Yourself. Some of you may be offended. But wait: this is not psychology, not psychobabble. No, unfortunately, this is part of the human condition. Each of us is a unique mix of strengths and weaknesses, carrying both the potential for greatness and the seeds of our own defeat.

Have you ever won a complex technical challenge, only to have victory sour from your ignoring the simple stuff such as forgetting to dot the project "i" or cross the technical "t"? That's typically how our strengths and weaknesses interact. This doesn't have to be: not if you understand your weaknesses and use mental jujitsu to push them off base and immobilize them.

Supposedly the ages 35 to 55 are considered the golden age for success. Why? Because at that age we stop fighting our childhood fears and battles, and come to terms with who we are, complete with muscles and warts. We can then build on our strengths to get things accomplished, while trapping or immobilizing our weaknesses to avoid self-induced slip-ups.

Dirty Harry, the movie rogue cop, said it best: "Every man should know his own limitations." He was right. Learn your weaknesses and then figure out how to work around them.

30 Job Search Tips

Overview

Fortune favors the prepared mind.
— Louis Pasteur

Technology is fun; job hunting ain't. No one likes the process, not you the candidate, nor the interviewing managers. This section can't change that. But it can help you be more effective, more focused and ultimately more successful at landing the right job.

Gathered here are a set of observations, concepts, and techniques that can help. You might not like what you find here, but it's not my purpose to entertain or beguile you. No, it's to ensure that you get the best job possible.

This isn't an everything-you-need-to-know guide to job hunting. Instead, these hints and observations go beyond the common knowledge pontificated in job hunting manuals and the popular press. This stuff works.

Catching a Horse

It's easier to catch a horse when you are riding one.
— Chinese proverb

It's also easier to find a job while you still have one. If you have a job and things are getting rocky, and your dander is rising — don't blow up; don't quit. Take a breath; sing "100 Bottles of Beer on a Wall," go for a walk, take a long run. Whatever you do, don't quit.

Back off. Try to analyze the problem and see if things are really as bad as you feel they are. Talk to a more experienced technical coworker or another manager. Don't be too hasty; there may be an alternative solution available.

If not, if it's best that you leave, take your time. Figure out what you want to do and start looking. If your job situation is critical, go talk to your management and make arrangements to leave. If you can't talk rationally with your immediate manager, then talk to his or her boss. Most companies will work with you to ease a difficult situation.

Contacting People

Reach out and touch someone.
   — AT&T commercial

While job hunting, one of the hardest things to do is to contact the people you need to talk to. And it's getting harder each year, what with the proliferation of electronic mail and automated answering systems.

Other than a lack of chemistry or qualifications, probably the number one reason most jobs fall through is a lack of continuous contact between candidate and hiring managers. Just as separation is hard on a romance, minimal or no contact between you and the hiring company generally leads to a slow decline in interest. If a manager wants to hire you, then it's up to you or your headhunter, if you have one, to maintain contact and not let the hiring situation die a slow death.

Here are a few hints on how to stay in touch:

1. Phone — it's very hard to catch busy people by phone. So call them before 8:00 A.M., over lunch, and after

5:30 P.M. — that's when busy people hide in their offices to get real work done. And that's also when phone calls usually are not filtered by secretaries. If you get an answering system leave only one message, don't clog up their voice mailbox. Just keep calling back until you get the person.

2. Electronic Mail — get the electronic mail addresses of the hiring managers (it's usually on their cards). And use it. In an interview, don't forget to ask for their e-mail numbers. Electronic mail services are available and are interlinked; you can get from one to the other. All you need is a PC with MCI or Compuserve. Use this judiciously. Have a reason for each message and try to be polite and entertaining, not a nag. Whatever else you do, don't refer to unanswered messages.

3. Cards — you can get managers' attention by sending them interesting greeting cards. Such cards can serve as eye catching and inexpensive reminders. Send something funny or colorful, but appropriate — hiring doesn't have to be a stuffy, dry affair. These days you can actually customize your cards at some greeting card shops and thus create an effective message.

4. Technical Material — another way to maintain contact, to make an impression, and to just be plain helpful to a fellow techie is to send supplemental technical material, usually on topics discussed during the interviews. Sending such material shows that you paid attention and reinforces your position as a technical professional.

5. FAXes — Don't forget FAXes. Generally people read their incoming FAXes before they tackle their mail — it's more immediate. So don't hesitate to FAX correspondence. You can get an inexpensive FAXboard for

your PC and FAX directly from home. Or you can sign up with telecommuting services such as Compuserve or MCI, which have FAX capabilities — you can send e-mail to a FAX. With some FAX systems you can add interesting graphics to your FAXes. Just remember to keep them appropriate.

Counteroffers

*Far and away the best prize that life offers is
the chance to work hard at work worth doing.*
— Theodore Roosevelt

A company is known by the people it keeps.
— Anonymous

Counteroffers are a tricky bit of business. Just as you view
your company as a family, so do many managers. And they
can feel betrayed, as only a family member can, when one
member — you — opts for divorce.

Unfortunately, in some organizations the only way you
can get management's attention focused on rewarding con-
tributors is to quit. So I won't say don't use looking for a job
as a lever to get a deserved raise or promotion. But I do say
be very, very careful in how you go about it. And be prepared
to leave if your ploy doesn't work.

In my experience I've found counteroffers to be counter-
productive (pun intended). Generally, going to the brink,
while effective at getting money or position, usually doesn't
solve the deep seated structural or personality problems that
started you down the good-bye path in the first place. Quitting

and then taking a counteroffer to stay seems to move most folks one step closer to an inevitable departure.

I'm not without sin; I too have taken a counteroffer, and it did work out rather well. However, in general I've not taken them, and have parted companies on the best of terms.

Finally, you have to be very careful not to give the impression that you interviewed for jobs as a cat's paw to get a raise or pry some goodies out of your employer. Remember, you just may need one of those jobs you interviewed for, especially if your new arrangement doesn't work out.

Do It in Parallel

Whatever worth doing at all is worth doing well.
— Phillip Dormer Stanhope (1694 – 1773)

If you need a new job, be sure to explore all opportunities in parallel. Don't try to search in a serial fashion, sequentially tackling one after another. You can't afford that approach.

Being laid off or forced to find another position can be frightening — there is no long-range security. We can lull ourselves into a false sense of security by having a long list of job possibilities. And then sequentially go through the list, one by one. Why? Because deep down inside we're scared that if we charge through this list in parallel and exhaust it, then there's nothing left to hope for.

Don't give in to this temptation. Get out there and do your job search in parallel. Run through the possibilities. There are always new opportunities, if not in one field then in others. The sooner you exhaust possibilities in one segment, the faster you can move on to another segment or category.

You probably don't have the resources or time to waste in fooling yourself. Go full bore. But remember that just because a list of possibilities didn't pan out, that doesn't mean that conditions won't change and one of these rejections may turn into an acceptance (see Handling No). That exhausted list may still be a potential resource.

Don't Get Mad

An angry man opens his mouth and shuts his eyes.
— Marcus Porcius Cato the elder (234 – 149 B.C.)

Don't get mad, so you can quit your job. For many of us, our companies are akin to an extended family. Walking away from that family can be traumatic, so traumatic that many folks must get mad at their managers or company just to be able to cut the cord and leave.

Don't get mad as an excuse to quit. Leaving your company can make rational sense: you've reached a point where you can't grow, or a new job at another company offers an incredible opportunity. There can be lots of solid reasons to leave. After all, at some point, we became adults and left the parental nest. Similarly, we can grow to the point where we can leave a company for better opportunities.

I've seen lots of technical people get a full head of steam up at their employer, when it made perfect sense to leave gracefully and take a far better position elsewhere. Getting mad can have a high cost: the person you yell at today, may

245

well be someone you work for or with downstream. Or worse, he or she can be the individual who blackballs you years later.

Finding a Job

Think before you act.
    — Aesop (620 – 560 B.C.)

Nobody likes looking for a job; it's a hard, demanding task —
and scary to boot. But if you want or have to look for a new
job, make an organized, professional effort.

Once you start looking, you can't afford to be sloppy and
leave out potential jobs. The job you take might not have
even come close in comparison to the one around the corner,
with new technology that you didn't even know about.

Take an organized approach. Do the following:

1. Characterize your own career and skills. Write a
   mini-resume listing jobs, skills, different technologies and tools you are familiar with.

2. What do you want to do? Make a list of the kind of
   jobs you lust for and the technical areas you want to
   get into. Make a comprehensive list (paper is cheap).

3. Where do you want to work? What are your geo-
   graphical options? Must you stay in the local area,

can you commute, would you consider moving, and if
so, where?

4. Who's out there? Given your skills, the type of work
    you want, and your geographical needs, what compa-
    nies meet your needs? Be thorough. Here are some
    hints:

    - Ask around, talk to people.
    - Look in your technical press, journals for compa-
      nies doing work you're interested in.
    - Look in the area newspapers. Remember each area
      has one dominant employment paper; find out
      which paper is dominant and use it.
    - Look at old newspapers, especially the Jubilee or
      special employment editions. These are generally
      in the main library on microfilm.

5. If you have a skill with a specialized software pack-
    age or with special hardware, don't hesitate to ask
    the software or hardware supplier for job clues. The
    product sales or marketing people may be able to put
    you in touch with other companies that use the hard-
    ware or software and thus need your specialized
    skills. Try to do this privately and, if they do help you
    out, keep the fact that they helped you private, espe-
    cially at your current job.

6. What jobs are there? Remember only 20% or 30% of
    all job openings are actually advertised. Look at the
    technology and products to identify companies that
    can use your skills.

    - Look in newspapers, employment and technical
      journals (don't limit your search to today's publica-
      tions, look at older issues too).
    - Talk to friends, acquaintances. See what is hap-
      pening out there. Get in touch with ex-coworkers,
      people you went to school with.

- If you wish, talk to one or more headhunters (see Using a Headhunter).

7. There may be many different job arenas that you want to work in. Write a resume for each. As you contact companies, you may find it necessary to tailor a resume for each company. Be prepared to do what works best. Always do a new cover letter for each company (see Writing a Resume).

8. Start sending resumes, contacting companies, calling folks referred to you by friends, and, if you wish, working with headhunters. Explore all these avenues concurrently.

9. As the interviewing process moves along, continue to look for new companies and jobs. Try to get a handful of good situations to choose from lined up at the same.

10. Get your offers and analyze the potential job situations. Make the best choice and notify the others of your decision (see Making a Decision).

11. Once you've found and accepted a new job, then quit your current one. Do it in a high level, professional way (see Quitting a Job).

Flight Versus Fight

Success is never final
Failure is never fatal,
It's courage that counts.
— Winston Churchill

The world has no room for cowards. We must all be
ready somehow to toil, to suffer, to die. And yours is
not the less noble because no drum beats before you
go out into your daily battlefields and no crowds shout
about your coming when you return from your daily
victory or defeat.
— Robert Louis Stevenson (1821 – 1845)

There seems to be a trend, an emerging pattern for a large number of technical professionals — declining employment stays at companies.

It's easy to spot. Many technical people work for 5 years or more at their first company, and then migrate to different companies, with each job stay declining, eventually dropping to periods of 1 to 1 1/2 years. Sometimes this churning through jobs can't be helped, especially in today's fast-turning job world where a lot of yesteryear's companies have bit the dust.

But something else seems to be happening: some learn that it's easier to go to another job, rather than to stay and slug it out at the current one. What they've acquired is a "flight over fight" behavior. And that tactic can be very destructive to a career. After all, companies hire you for your integrity, stick-to-itiveness, and technical talent. If you can't hang in there and fight the good fight, why should someone else trust you, let alone hire you?

If you're looking for a job because things are a bit hot where you work, stop and make a simple assessment. Back off and look at your past job history and ask yourself if you've taken to the job hopping trail. If so, toughen up and learn to hang in there and duke it out. If the good people won't stick and fight, then a lot of projects, even companies, will eventually go down the drain.

And beware of dropping out just before your project moves into implementation or the product home stretch. Leaving at that juncture shows a lack of commitment. Your project will suffer, your reputation will suffer, and your resume will suffer. Managers don't look at such timing with favor — they know what it costs and what it could cost them. There are times when the project needs take precedence over individual needs.

Get Excited

Twixt the optimist and the pessimist
The difference is droll;
The optimist sees the doughnut
BUt the pessimist sees the hole.
— McLandburgh Wiilson
(Pessimist and Optimist)

Everybody hates job hunting. It's messy; it's grueling; it's frightening, and it has long-term, high-stakes consequences. But this is your technical life — not a dress rehearsal. What happens now is for real. So you might as well relax and enjoy the trip; it's the only one you'll ever have. Get excited, technology is fun stuff.

All things being equal, a confident, buoyant and turned-on candidate will be hired over a quieter, more restrained and less assertive person. Managers, as well as most people, like turned-on people: they are easier to manage and usually turn to with a vengeance. General Colin Powell, who headed the United States Joint Chiefs of Staff, said it best: "Perpetual optimism is a force multiplier." ** Optimism brings its own rewards.

I'm not saying that you should change your personality and be someone you're not. And if you are a super-programmer or super-system-designer, hey, it won't matter. But if you are one of the rest of us, you might as well take a positive view toward interviewing and enjoy it.

Think of it this way: you get a chance to see the insides of a lot of projects, evaluate some high tech stuff and — best of all — you might learn something useful.

** *The Private Powell,* by David Roth, September 3-5, 1993. USA Weekend, Arlington, VA.

Getting Fired

The test of success is not what you do when you are on top.
Success is how high you bounce when you hit bottom.
— General George S. Patton, U.S. Army

Getting fired or laid off is an increasing possibility in today's high-tech world. It's becoming common in the 1990s and is something that you had best be prepared for.

There have always been companies, such as aerospace firms, that have dramatic manpower shifts depending on orders or contracts. But today that instability is more widespread as many of today's larger firms are downsizing, dropping down to fighting weight. And the aerospace world is in a long-term decline as well. Moreover, the marketplace moves so fast that many organizations simply lose out. So the possibility of losing your job is not an abstraction any more —it's an increasing reality.

When you're fired or laid off, here are some things to do:

1. Don't, I repeat, don't get mad. Stay calm. If you have to, ask for a little time to think about what's happening before you discuss conditions and settlements.

It's important that you be in control of yourself, on top of your form, when you talk over your termination details. There will be no second chance. Take the time you need to be in control of yourself.

2. Don't take it personally, no matter what the circumstances are. Be professional.

3. Most organizations and managers are not very good at layoffs or dismissing people. Many botch it. That's more a sign of their inexperience and discomfort at the task, than disrespect to you. So don't be insulted if it's not handled well. Use it to your advantage.

4. Unless you have another job on tap, try to delay the termination to give you time to find another job. Stay on as long as possible to give the appearance of employment as a base for your job search.

5. If you have to leave immediately, ask for phone, copier and secretarial privileges. Try to have calls fielded for you and relayed to you for a period of time.

6. Revise your resume and start looking for a job right away. Don't take a few weeks off to lick your wounds or to try and forget the situation you're in. Go after a new job right away.

7. Find out why you were fired or laid off. Don't argue; just find out what their reasoning was. If it was extremely unfair or wrongly done, be sure to consult with someone before you make any accusations. Be sure of your ground for any actions; and even then think twice about taking any legal action. Such actions can take a long time and can literally drain both your emotions and creative drive.

8. If you can, finish your work, update and turn over your files to whoever inherits them. Don't just dump it on the floor and stomp out. Take the high road on

your way out; people will remember and think well of you for it.

9. Stop and change your thinking. The company's problems are no longer yours. The cord tying you to the company is cut; now is the time to put yourself first.

10. As hard as it is to do, be philosophical. Believe it or not, things tend to work out for the best. The world doesn't end. I should know — I've been fired and laid off, and each time I ended up better off.

11. If it's any consolation, staying on in an organization going through traumatic layoffs and bad times ain't fun. It can be downright awful. Going to another company that's on a growth swing is a lot more pleasant than hanging on, watching a company eat its young, and wondering who's the next to go into the pot.

12. Don't look back. Don't go through any agonizing what-ifs. The past is a sunk cost; you can't change it. All you can do is change the future.

13. A lot of good people, strong productive people who put in the time to get the job done, end up getting canned. It may not be fair, but it happens. Don't feel betrayed. You built, you did, you accomplished, you grew, and you're the better and the stronger for it. Now take your strengths elsewhere.

14. Make sure that you get the names, addresses, and phone numbers of the mangers and coworkers who like and respect you. You'll need them as references and perhaps for job leads.

15. As you're phasing out at the company, don't be shy in asking others if they know of any outside jobs. Or if they know anybody who might know or who you might contact to find out. People will want to help you if you give them a chance.

I found out about layoffs early. As a newly minted EE from college I went to Librascope (Glendale, CA) to write logic-level computer diagnostics. After six months there was a layoff. We were gathered into two groups and called into meetings. I was in the first group — those who were told they were valuable and would stay. I was just starting to feel safe, when the manager stopped his speech, looked at me and said: "Oops. Ray, I think you're in the wrong group!" Not an auspicious moment.

But I wasn't terminated; instead, they assigned me the gruesome task of writing Naval ordnance alteration reports (OrdAlts) for torpedoes. It was awful. I stood it for three days and came in early one morning to quit. I couldn't find my manager, his boss or even his boss's boss. My office mates were in stitches as I doggedly searched for someone to accept my resignation.

I worked my way up the chain until I found the chief engineer, Hank Buchard. I thanked him for the job, but said I hadn't become an engineer to push paper; I'd quit first — I wouldn't even finish the stuff on my desk. Hank was pretty good about it; he set me to designing logic front ends for large (then) commercial disk drives. I learned two lessons: (1) even in layoffs things can work out; and (2) it pays to be true to your goals.

Handling No

Is this a real NO?
— Salesman's response to a turndown

Everybody hates rejection; even top salespeople can go cata-tonic from too many NO's. If you go out and interview for jobs, invariably you're going to collect some NO's.

Make the best of a bad situation; use turndowns to your advantage. Take the high road, be professional, and try to understand the situation from the hiring manager's position (if you think that you're uncomfortable being turned down, just think how they feel saying No).

Then again, a NO might not be a real NO: the hiring sit-uation may have changed; the rejection decision might have been someone else's; they may have had to accept an internal transfer instead of hiring in. There are a lot of factors you cannot know about. Today's NO may very well translate into tomorrow's YES.

Listen carefully when they say NO. Ask questions. Is it a solid NO? Or did circumstances change? Should you talk to them later? If they liked you and it didn't work out; can they recommend an opportunity elsewhere in their company or at some other firm?

Make sure that the rejection is on solid grounds. Did some factor worry them and they had insufficient information to go forward, and then simply made a blind decision? Don't argue; don't get mad. If the decision can't be changed, don't push; leave it up to them. If possible make an appointment to go back in and clear up any misunderstandings that were hiring barriers.

Headroom

All Hell needs is water and a few good people.
— Texas pioneer proverb

Is there enough headroom for you to grow in this new job? Or are you hiring into a situation where your technical skills can be stymied and stunted?

Assuming that you're not desperate, consider the headroom question carefully. You don't want to be frustrated by a lack of technical, product, or managerial growth.

Are there significant things to learn, skills to be had in this company? Will you acquire new skills? For example, as a programmer will you work in a new operating system environment, say Windows NT? Or as an engineer, will you design in different microcontrollers? Will applications and products require substantial learning on your part? Given the current technology and the company's methodology, will there be significant technical growth in the company and the product technology?

As a junior or intermediate don't hire into a job where you cannot either acquire or beef up technical, product, or managerial skills or responsibilities. However, if you've turned the skills corner, where you are looking for a place to practice your current expertise — can you do that in this job? Can you effectively contribute at your expertise level? Will they listen to you and utilize your skills?

Interviewing 1 — Basics

Men don't plan to fail, they fail to plan.
  — William J. Siegel, Printz-Biederman Manufacturing

Few great men could pass personnel.
  — Paul Goodman

1. Do basic research on the company before your interview. Look them up in technical publications, and the trade press. Talk to friends who work there or know about the company.

2. Be on time. Get there early; show up at least 5 minutes before the interview. If you will be late, call and reschedule the interview.

3. Dress up. The general rule is dress up one or two levels above your working dress. For you wild programmer and engineer types: you're not kowtowing to the powers that be. You're simply demonstrating that you're capable of dressing up as the situation warrants, and that they won't have to hide you in some back room when customers come calling.

4. Be polite to everybody, especially receptionists and secretaries.

5. Relax and be cheerful. You really have nothing to lose: after all you don't have the job, so have a good time.

6. Don't get in arguments, especially technical arguments, with the interviewers. From your viewpoint this is an exploratory session; later, if they make you an offer, you can get into more detailed discussions.

7. Bring a small note pad for notes. Get business cards from all interviewers; if they don't have cards, write their names and titles down. Try to get E-mail, phone numbers, and FAX numbers.

8. Bring 4 or 5 copies of your resume with you. Don't assume that they will have it, even if you sent your resume before the interview.

9. Bring samples of your design work including documentation, schematics, or code. But be sure that these samples don't reveal any secrets from your current employers. Try not to leave these samples or let them make copies. If they insist make sure they get only small, unimportant samples of your code or design. Remember, that work belongs to your employer, not you.

10. Be prepared to discuss your design philosophy and methods, as well as details of your projects. But don't feel compelled to give competitive information on your current employer's products or technology. Use common sense here.

11. If you sit at a table with a number of interviewers, you can keep track of them by placing their business cards in front of you to reflect their seating order.

12. Remember, they are taking the time to talk with you because they have a need.

13. At the proper time ask about the project, the company, and (if you don't know) the industry. Ask where

they are headed, technology- and product-wise. Your research can help you here.

And finally, be prepared for the unexpected in interviews. Strange things happen. One of the oddest interview stories concerns John von Neumann, one of the intellectual fathers of the modern computer.

Von Neumann held interviews for a chief engineer position on a pioneering computer development project at Princeton. Julian H. Bigelow interviewed at von Neumann's house (he got the job). When von Neumann opened his door, a dog entered the house with Bigelow. During the interview, the dog romped about making a general pest of itself.

As Bigelow was leaving, von Neumann politely asked if he always traveled with a dog. That's when both discovered that neither owned the dog — it had simply wandered in to explore a strange house.

Interviewing 2 — Priorities

You can't think and hit the ball at the same time.
— Yogi Berra, baseball player, manager, coach

Keep your priorities straight. Remember the purpose of an interview is for the interviewer to determine whether or not to hire you.

An interview is not a place to discuss opinions, to brag about technical skill, to talk about your life, or to detail your wants and needs. Rather, it's the place to make a clear and concise case of why the company should hire you. It's that simple. Do anything else and you'll be wasting both your and the interviewers' time. And especially don't whine about your current job and managers. Today's dissatisfied employee has a good chance of being tomorrow's dissatisfied employee.

Remember, the interviewing managers are talking to you for one reason: they need technical manpower. And what they need to know is: can you do the job; do you fit in? Are you a good long-term investment?

Bringing in extraneous matters, especially ones that introduce doubt in their minds — such as, "I'm not sure that I want to do this," or "the commute may be a problem," or "the task seems undefined" — just clouds the process. Let them decide on you first.

To repeat: It's your job in an interview to convince interviewers that they need to hire you. Let them make that decision first, then get into discussing other adjustments. Look at interviewing as a two stage process: one, they decide that you are the person they need; and two, then you decide whether or not their offer is the right one for you. Don't confuse the two stages. Satisfy interviewers first, then satisfy yourself.

This advice holds, by the way, in both good times and bad times. In good times, you might not take the job, but you've convinced some people that you're good — that perception can come in handy later. And in bad times, you've optimized your chances.

Interviewing 3 —
Selling

Words are, of course, the most powerful drug used by mankind.
— Rudyard Kipling (1865 – 1936)

Use an interview as an opportunity to sell your capabilities.

You are in an interview for one reason only — to sell the interviewers on you enough for them to offer a job. The good news is that you are there, the bad news is that there is no guarantee that they will hire you or even hear the reasons why they should.

It's up to you to focus the interview, to give interviewers the reasons that they need to hire you on. Not only must you answer the questions that they ask, and establish your credentials, but you need to focus the answers toward satisfying the major question of hiring you or not.

I'm not advocating playing games or lying (never), but rather refining your comments and answers to help the interviewers make up their minds. The idea is to help them, not con them. Moreover, they may need help, many managers are not very good at handling interviews.

Use every opportunity to show why they should pick you. For example, suppose the interviewer asks if you know a specific computer language or logic family. If you do, then say yes and expand on what you know and how it applies to what they are doing. On the other hand, if you don't know it, point out it's similar to one you do know, and expand on how that will help you in their job.

Don't just sit there like a cigar store statue and mumble monosyllabic answers. Ask what the proposed job entails, and show how it dovetails with your prior experience.

Interviewing 4— Closing

It's all for nothing if you don't ask for the sale
— Sales proverb

The essence of sales lies in clinching or closing the sale. Convincing somebody to buy something does no good if in the end you don't ask for the order. The same thing is true of an interview — you have to close it productively.

1. If you are interested in the job, tell them so. You don't have to say that you will take the job as is, but that you are definitely interested in pursuing it.

2. By the same token, if you are in love with the job, the people, and the company, tell them so. If you really want it, you can commit to it, but leave in a loophole — "provided we can work out the details, and I see no reason why we can't."

3. After you express your interest, ask them what the next step will be. Do you come back for more interviews? When can you receive a preliminary decision? What is their hiring time frame?

4. Try to get a solid action item: They or you will call by a certain date. Don't leave things hanging.

5. If they won't commit to letting you know in a relatively short time, tell them you'll call and check in.

6. I'm not a big proponent of follow-up letters — they all sound so depressingly similar, chock-full of cliches. However, if this is a formal company it's probably a good idea. For most interviews I favor sending in something germane, but humorous. I used to go out and find appropriate greeting cards and send them as reminders.

7. If you went in through a headhunter or the personnel department, don't hesitate to call them later and check up on any progress. Be sure to ask if there are any difficulties that you can straighten out — you never know what can pop up. (In one case, a manager almost didn't get hired because he used his boss as a reference and the hiring managers were unsure if he was being laid off or not. It turned out that the boss was a friend of his — they hired him. It pays to ask.)

8. Don't look on the final job negotiations as a contest with a winner and a loser — everybody should win. And don't wait for them to make the first move to reach an accord; be willing to compromise and move things forward.

Interviewing 5—
Have Fun

He who laughs, lasts.
— Anonymous

Have fun. Don't take interviewing too seriously. You don't have the job yet, so you've nothing to lose. Relax, enjoy yourself.

While interviewing you walk a fine line between formality and informality. On one hand, it is a formal situation and the interviewers want to see how you deal with it. On the other hand, if you're stiff and formal, you won't come off as someone who easily fits into the people part of their organization.

Taking interviewing too seriously can cause you to choke up and be ineffective. Yes, it's a tense situation, but remember you're also visiting technical colleagues. Make your points about your technical skills, your drive, and long-range technical goals. But why not be relaxed and do so graciously?

Not only that, but the interviewing manager(s) may be as or even more uncomfortable at interviewing as you. Many managers don't like interviewing and are not very good at it. If you make it easier for them, they'll love you for it.

Remember, you're a technical professional, as are the interviewers — you belong to the same club, and have paid the same dues. Over the years, you'll run into each other again. So if things don't pan out today, they may do so tomorrow. Consequently, take a high-level tack; these people may end up as part of your technical future.

Look before You Leap

Think before you act.
— Aesop (620 – 560 B.C.)

There's few things as uncommon as common sense.
— Frank McKinney "Kin"Hubbard

Before you take a job, make sure that there will be a job there in the future. Check out the company, and division financials, as well as common knowledge on the group's product line. Talk to the technical people you know, as well as professionals such as stockbrokers. Ask.

These days the dividing line between survivor companies and dying companies is very fine. Be sure that the company, the division, and group that you're planning on working for will survive. Feelings and technology are not enough. Check the company's stock, financial expectations, as well as the health of the division or group that you'll work in.

How well do(es) their product line(s) do in the marketplace? Are they vulnerable? Is the company willing to commit resources to its product line and division? These are all questions that need positive answers or you may well be leaping from today's frying pan into tomorrow's fire.

Interestingly enough, one study (*The Corporate Steeple-chase* - see Bibliography) found that most people — even those who played the stock market — do not do a thorough check of a new company. Don't be one of them. Be aware of what you are walking into. Just because a company is doing well, that doesn't mean that a specific division or product line is also doing well. Check it out. Better safe than sorry.

Money

We work to become, not to acquire.
— Elbert Hubbard

We're overpaying him but he's worth it.
— Samuel Goldwyn (1882 — 1974), movie mogul

For a society that uses money as a scorecard, we're particularly uncomfortable talking about it. Doctors are above it, engineers are not supposed to discuss it among themselves, managers hate allocating it as raises, and none of us like dealing with it in job negotiations. Even so, salary is something that you must deal with in changing jobs.

Before you negotiate salary for a job offer, try and find out where your salary rates against industry scale for professionals with your degree, experience, skills, and time in grade. Are you underpaid? Are your overpaid? What can you reasonably expect in salary and other benefits?

There are a number of factors that you should consider in setting your monetary expectations for a new job. These include:

1. Money is not the key factor, long-term survival is. Needed are the skills and the experience that the marketplace will value.

2. Salaries generally are not set by hiring managers. They must follow company rules and guidelines that limit raises and set salary boundaries for each employee category.

3. Salaries are unfair. People are not generally paid for what they do. Instead they are paid based on their salary history, grade, and job history. Those that job hop tend to have higher salaries, as do professionals from higher cost sections of the country.

4. Take a look at the hiring company's job categories and the corresponding salary grades. You don't want to be hired in the top of an existing salary grade, unless you are fairly sure of promotion up to the next grade. Otherwise you can top out in the grade and end up with limited raises over the next few years.

5. In general you get your large salary increases by moving to another company. Few people get significant raises from their current employer because company rules generally prohibit them, even if well deserved. Consequentially, most managers cannot correct gross salary inequities easily.

6. You can negotiate for more than just salary. Other items worth pushing for are job titles, special insurance, perks, off-hours, and a car.

7. Check into other benefits including medical, dental, and life insurance. A large raise can vanish if other costs rise (larger insurance premiums or a need for supplemental insurance).

8. Is there a 401K to help you shelter income for retirement. Is it portable?

9. Don't make "I" demands — "I want, I need, I must have..." Don't make it personal or a test of your personal worth.

10. Find out what the company's raise and review policies are. What are the review periods for your job category and what are the typical raises or the raise pool? How do they rate people for raises? Is there a salary freeze? Is one coming?

11. What is their promotion policy? Where would you fit in that? What career path do they see for you, what are your options?

Remember your priorities. You don't want these compensation negotiations to sour your future working relationships. Be reasonable and work toward a common solution. There is no point in negotiating a high salary to work for people who you've offended.

Networking

Nobody knows you when you're down and out.
— Anonymous

I really hate the term networking. I get this mental picture of someone, out of work, mindlessly calling up everybody he or she remotely knows or ever bumped into. That word is now even used as a verb — as in "to network."

All of us, managers, programmers, design engineers, technical professionals, belong to a technical fraternity/sorority, a community of doers. We work with each other, compete against others in different companies, and learn from each other.

As members of a common community we should talk with and help each other. This community is your network; and you should be working with it *before* anything happens that puts you on the street.

Be active in this community. Go to seminars, take classes. Read articles in the technical publications and trade press, as well as on Internet or Compuserve forums. Don't be afraid to call up or message an author who has written something helpful in your work. It's only fair, in times of need, to use this net, this community of fellow designer/managers to help you out (you today, me tomorrow).

Don't be afraid to join technical societies and serve in some capacity. Most technical disciplines have conferences and seminars, usually put on by technical societies. See if you can help out there.

Obviously when you're out of work is not exactly the best time to discover that you're a member of a technical community. Talk to people now, when you can, not later when it's too late.

No Sure Thing

Two choices is ten times better than one.
— John Choisser, Annabooks

In a job search there is no sure thing. No job is ever in the bag. Too many things can go wrong to ever be able to count on any one hiring situation.

Therefore, until you have an actual job offer in hand, don't count on any specific job situation coming to fruition. A lot of mishaps can happen from interview to offer: the firm implements a hiring freeze; they may bring in somebody from another office; the job description may change; the project can get canceled; or the hiring manager can get hit by a beer truck.

So once you start looking for a job, keep at it until you have enough offers to choose from (see Do It in Parallel). For the same reason, be careful in your commitments. When you commit to a job in the early phases before an offer, do it with a time loophole. Say things on the order of: I really want this job, and if we can get together on the details in a reasonable time I'll be here.

And be calm when things start to crumble and fall though. Assume that the folks involved did the best that they could. And thank them for it. There will always be time for another try. Keep your cool.

One thing more; you don't have a legitimate job offer until you have it in writing. Don't quit or turn off other offers until you have a written offer.

Picking the Best Job

The rarest gift the Gods give to a man is the capacity for decision
— General George C. Marshall, U.S. Army

So here you are lucky enough to be holding a number of job offers — now what? You can't stall forever. How do you know which job is best? Should you go with your instincts? Or should you pick the one that pays the most? Or the one that is the most exciting?

Here's a quick and dirty way to make your choice. Make a matrix with the job choices on one axis and a list of key factors on the other. These factors can range from ease of commuting to an evaluation of the company's technology base. Others include salary and perks, management quality, product market position, technology value of the skills developed there, product interest, development resources, work environment, and quality of employees.

If building the table entries opens up new questions, don't hesitate to call and get more details. Just do the full analysis first, and then make your calls. Get all your questions together first so you can apply them to all the offers.

Go through and grade the offers, giving each company's factors a score from 1 to 10, with 10 as the highest grade. When done, review the table and make any adjustments you need to make the scores consistent. Now total up the scores for each job and compare them. Spend some time reviewing the scores and their meanings.

Now toss the score sheet away and make a decision based on your gut feeling and perceptions. This technique combines the best of two worlds: a detailed analysis of job selection factors, and your own sharpened perceptions. By analyzing the factors first, your perceptions and views are expanded and sharpened. But no table, no set of factors can define the human spirit and drive needed for success. That's why the final selection should be from the heart (or gut).

Quitting in Style

Parting is such sweet sorrow.
— William Shakespeare (1564 – 1616)
(Romeo and Juliet)

Forget the bad times and stress the good ones. End your job on a high note, not a sour one.

Remember no segment of the technical world is truly isolated. Most companies are interlocked, tied together via game trails that people move over as they shift from company to company. So don't be surprised if the people you worked with at your old job turn up later at some organization that you want to work for.

Leaving a company can be stressful. Make it as pleasant as you can, both for yourself and the others. You can afford to be magnanimous, you're probably leaving an old, stagnant situation and moving on to a better one.

Finally, you've shared some good times and hard work with your old cohorts and managers. Celebrate common successes and wish them well (and mean it).

Here are a few more hints:

1. Try to give positive career reasons or objectives for leaving.

2. Keep away from the topic of money as any kind of driver for your departure.

3. Don't hide where you are going. Be up front about it. If you do hide it, that will make you appear to be paranoid, neurotic, or just plain mean-spirited.

4. Never mention personalities as a factor in your decision to leave (either when quitting or at your new job).

5. Stress what you have learned, not the problems.

6. Don't hold any grudges. Be open, friendly, and aboveboard.

7. Wish them well and mean it.

8. Have a good time in your remaining employment. Whatever the problem(s) were they are now irrelevant. Do what needs doing and relax.

Simpatico

Good people attract other good people.
— Roberto Ruiz, Maya Construction Co.

A job is not just a job. Are these guys (and gals) at the new company simpatico? Do you share a technical excitement, a common technical view, and a collective methodology? Are your priorities the same?

If you don't share these things, working there can be more difficult than you think. Are they more advanced than you and do you want to acquire their technical viewpoint and methodologies? If so, that's fine.

But if you're coming in with more advanced experience and skills, then what? Do you think that you can convince them to move to your side in how to design and do things? Can you live there if you cannot convince them? Can you operate with less professional effectiveness than you're used to?

If and when the company makes you an offer, that is the time to work out some of these issues. You do not want to be a square technical peg in a lower-level round hole. Talk to them; you may be surprised how flexible they are. But remember, almost no one likes a new hire to come in and start criticizing everything. Be gentle, take your time, make your points, and you'll probably win.

Small Versus Large

One man's utopia is another man's manure pile.
— Graffiti, downtown San Francisco

There is no security on this earth; there is only opportunity.
— General Douglas MacArthur, U.S. Army

The differences between working for a small company versus working for a large company used to be pretty clear cut. Small companies were highly reactive; you wore a lot of hats and did a range of things; and worked in a fairly free-form environment. In contrast, large companies were much more structured; you generally worked in a narrow skill area, and had lots of picayune work rules and procedures. And big companies generally paid better than small ones.

In years past, working for small companies was supposed to be risky, while working for Fortune 500 companies was a guaranteed ticket to a slow ride to retirement. No more. Fortune 500 companies are shedding middle management and technical talent, as easily as a snake sheds its skin. In the 1990s, large companies are job losers. Conversely, small companies are growing and adding jobs. In reality, there's no real, long-range security anywhere for employees of either large or small companies. There are no guarantees that any company will be able to stay ahead of the technical product curve. Today's market heroes can easily be tomorrow's losers.

Moreover, the operational differences between large and small companies are shrinking rapidly. Big companies, the mastodons of the 1980s now have fast reaction development teams. For example, take HP's team, which developed a 1.8" PC hard disk drive in record time.

In choosing between large and small companies, look at their resources. Does the company have adequate resources to enable you to do your job? Does it have the equipment, plant, and people needed to get the work done? Does it have the market resources to move your products to the marketplace? And finally, does it have the resources and will to stay in the market?

It's no longer true that big companies always pay better than small ones. A number of small companies actually pay better than large ones. Why? They hire the best and demand results. However, pay and benefits are still higher at large rather than small firms. But don't count on it.

And finally, personalities are more crucial in small companies than in large ones. It's critical that your personality mesh with management's personalities, especially in very small companies. Typically, such companies cannot handle too much dissension; they don't have the resources for diversity. So be sure that you and management are on the same wavelength.

With larger companies it's not so much the personalities of the managers that matters as much as the "operational" personality of the group, division, or company —the how and why they do things. If you don't mesh with that group's personality, you could have some major difficulties. You can come in and help them become more effective, but you had best be prepared for a long campaign.

Starting a Job

Turning over a new leaf is always fun.
— Anonymous

New roads, new ruts.
— G.K. Chesterton (1874 – 1936)

Habits are at first cobwebs, then cables.
— Spanish proverb.

Starting a new job presents an opportunity for you to get better. A new job means a new environment, new things to do, new responsibilities, and a new habitat.

Here's a chance to change your current work habits. As you adapt to the new environment, you take the opportunity to actually change, or to reprogram your habits. Put new, more effective habits in place as you acclimate to your new environment.

Don't get comfortable too fast. In starting at a new company, you will establish a network of relationships with the people there. At first you'll have a lot of time to work: you're not yet part of the schmoozing circuit. If you're smart you'll take advantage of that isolation and push out a lot of work first. It won't last.

Even better, you're starting out with a clean slate — a clean office or cube. You're in a state of, as they say, clutterless grace. Use it to your advantage and set up a clutterless environment (see Clutter).

Start on a productive basis, set your professional terms. Get an assignment, a task, as soon as possible, one with as much responsibility as you can handle. And finally, whatever you say you're going to do, make sure that you do it.

This is the best time to ensure that the interviewing agreements are met and honored. It's also where you can iron out potential problems before you sink into the day-to-day fray. Be professional and ask for what is best not just for you, but for the project and organization. You still have some maneuvering room to iron out difficulties. Use it.

Taking a Job

The best bargain benefits all.
— Anonymous

Accepting a job is where things get real, where the rubber meets the road. It is here where all the problems and differences between you and the new company need to be ironed out.

Up until now, you were concerned with convincing the interviewers to hire you to solve their problems. Now, at this stage you have to turn around and work together with them to solve *your* problems. Here is where you fix the details in place.

Never again will you be in as strong a position to set the conditions for your work as now. Here is where you do the technical horse-trading. This is where you set the stake in the ground defining your levels of responsibility, freedom, and accountability, as well as the areas and technologies that you will work in. Make sure that these commitments are understood — and be prepared to live by them, as well as to nudge others to do so as well.

End your acceptance on a positive note. They've just made an expensive investment and are embarking on a new professional relationship. Make them feel that they have made a good deal.

Oh yes, after painstakingly setting up your future work conditions, things can still change. It's not that unusual for you to start work on a totally different project with a changed reporting structure. In this technology business things change. If this happens to you and the reasons for the changeover are pressing, then go along with it. But make sure that your core agreements are recognized and honored. Be cool.

Technology Crapshoot

Technology will provide.
— The author

No one knows what he can do until he tries.
— Publilius Syrus (1st century B.C.)
(Maxim #786)

What you do defines what work is open to you. The technology you start in can end up being the technology you stay in. So be careful of what you work in: today's fun field can turn into tomorrow's pigeonhole.

More and more companies are unwilling to train good technical people in new technologies. Many will only hire those who have the requisite skills; thus, they end up competing for a small pool of technical folk. Gone are the days when a technical degree gave you license to lateral over into a new technology and pick it up.

Yes, you can change disciplines; you can change technologies, but it's not easy. You must take the lead, and put in the time and energy to pick up new design skills and capabilities. Read the literature, devour key technical texts, and work on your own at actually doing some of the new design work. Take classes and seminars, attend conferences, and learn whenever and wherever you can.

When you've picked up enough of these new skills, look around in your company. If there's a group doing what you lust for, go talk to them. Get some hints and pointers; hang around and help them where you can. When they need somebody you just might get a shot at it. All said and done, managers like to hire people that really want to be there and who love the work.